Future Resources and
World Development

FRONTIERS IN HUMAN ECOLOGY

Series Editors: **Paul Rogers**
The Polytechnic, Huddersfield, England

Kenneth A. Dahlberg
Western Michigan University, Kalamazoo, Michigan

J. Owen Jones
Commonwealth Bureau of Agricultural Economics,
Oxford, England

A Continuation Order Plan is available for this series. A continuation order will bring delivery of each new volume immediately upon publication. Volumes are billed only upon actual shipment. For further information please contact the publisher.

Future Resources and World Development

Edited by

Paul Rogers
The Polytechnic, Huddersfield, England

In association with Anthony Vann
The Polytechnic, Huddersfield, England

PLENUM PRESS · NEW YORK AND LONDON

Library of Congress Cataloging in Publication Data

Main entry under title:

Future resources and world development.

(Frontiers in human ecology)
Includes bibliographies and index.
1. Natural resources—Addresses, essays, lecutres. 2. Economic development—Addresses, essays, lectures. 3. Underdeveloped areas—Economic policy—Addresses, essays, lectures. I. Rogers, Paul. II. Vann, Anthony. III. Title.

| HC55.F87 | 333.7 | 75-46615 |

ISBN 0-306-30913-0

© 1976 Plenum Press, New York
A Division of Plenum Publishing Corporation
227 West 17th Street, New York, N.Y. 10011

United Kingdom edition published by Plenum Press, London
A Division of Plenum Publishing Company, Ltd.
Davis House (4th Floor), 8 Scrubs Lane, Harlesden, London NW10 6SE, England

CONTRIBUTORS

Dr F E Banks

Research Fellow in Economics and Econometrics
University of Uppsala
Uppsala
Sweden

Dr Biplab Dasgupta

The Institute of Development Studies
University of Sussex
Falmer
Brighton
Sussex BN1 9RE

Robert Dickson

W.D.M.
15 Kelso Road
Leeds
West Yorkshire LS2 9PR

Ken Laidlaw

W.D.M.
25 Bedford Chambers
Covent Garden
London WC2E 8HA

Dr Paul Rogers

The Polytechnic
Huddersfield
West Yorkshire HD1 3DH

Anthony Vann

The Polytechnic
Huddersfield
West Yorkshire HD1 3DH

PREFACE

Human ecology, the study of the interrelationships between man and his environment, is necessarily in interdisciplinary study. An aim of this series is to demonstrate the importance of appreciating the viewpoints of different disciplines on problems of the human environment when attempting solutions to those problems. It also aims to emphasise the crucial importance of integrating these viewpoints.

Nowhere is this more important than in the study of global resource use and prospects for international development. Early in 1973 we organised a symposium on the subject of "Human Ecology and World Development", the proceedings of which formed the first volume in the present series. That symposium was held within a year of two major international conferences, the third session of the UN Conference on Trade and Development (UNCTAD 3) in Santiago, Chile, in May 1972, and the UN Conference on the Human Environment held in Stockholm in June of that year.

The Stockholm conference marked the peak of a period of intense interest in environmental matters in many developed industrialised countries, an interest that had, initially, been concerned with problems of environmental degradation resulting from industrial activity but broadened out into an interest in problems of world resource use. It did not, however, extend in any great way to a consideration of the problems of the human environment in the less developed countries of the world. For the environmental movement, the main consideration was with the environment problems of the rich countries.

This was singularly evident by the almost total lack of concern by environmentalists over the results of the third session of UNCTAD. UNCTAD 3 had been seen by some observers as a final attempt to achieve cooperation between rich and poor countries in overcoming the problem of world poverty. The attempt failed, yet few people in the developed industrialised countries appreciated this.

We can now begin to recognise the long term significance of these two conferences. UNCTAD 3 demonstrated the lack of interest of the developed industrialised countries in genuine international development. The Stockholm conference helped sow the seeds of a new global phenomenon, the developing power of the producers of raw materials, by emphasising the global limitations on human activity.

Early in 1973, it was becoming obvious that the limitations which a finite global system were likely to impose on human activity called for a re-appraisal of our attitudes towards world development. If the less developed countries were to overcome the problems of human poverty which they experienced, there was a real possibility that this would have to be at *the expense of* standards of living being maintained in the developed industrialised countries. On top of this, the very fact that the poorer countries of the world were mainly producers of raw materials for the developed industrialised nations gave them an inherent "producer power"

At the time of the symposium on "Human Ecology and World Development" in April 1973, that producer power was just starting to be exercised by the oil producers. Now, some 30 months later, the process has developed faster than anyone would then have dared predict. So much so that possible long term trends in resource use in relation to world development are now beginning to get the attention they have so long been denied.

The aim of this book is to examine some of the longer term implications of the use of global resources on prospects for world development and on relations between the developed and less developed countries of the world. In order to do this, the book has been divided up into two parts.

Part I consists of papers prepared especially for this volume. It commences with a contribution from Anthony Vann on "The Global Ecosystem and Human Activity" which gives an ecological perspective to trends in energy and mineral resource utilisation. This is followed by a detailed case study of the one dominating example of producer power, oil. This is entitled "Oil Prices, OPEC and the Poor Oil Consuming Countries" and is by Dr Biplab Dasgupta. Following this, Paul Rogers contributes a paper on "The Role of Less Developed Countries in World Resource Use" which includes a consideration of the potential development of producer power for a number of primary commodities. This is followed by a paper by Dr F E Banks entitled "Problems of Mineral Supply" which concentrates attention on two important minerals, copper and aluminium. Robert Dickson and Paul Rogers then contribute a discussion on the possible effects of producer power on relations between developed and less developed countries. The editors then conclude with a summary of the views expressed in the preceding papers placed in the context of recent developments on the international scene.

Part II of this book consists of a number of important documents, statements and speeches which appeared over the period September 1973 to May 1975. This was the crucial period for the development of producer power and the contributions help to show exactly how the process has developed.

This is not to say that we are concerning ourselves with short term variations in policies and processes, rather that the documents included in this section have profound long term implications. We are considering here a phenomenon of recent occurrence, at least on an international scale, and emanating from less developed countries. If producer power does become a dominant force on international relations, and that now seems well nigh certain, then we must think in terms of decades rather than years in attempting to visualise its future evolution. Nevertheless it is a phenomenon which has basic implications for people throughout the world and deserves immediate and serious study. We hope this book, which attempts to bring together the approaches of several disciplines, will aid that study.

ACKNOWLEDGEMENTS

We would first like to thank Dr Dasgupta, Dr Banks and Mr Dickson for helping to ensure the completion of this book in little more than six months. We would also like to thank the authorities and institutions who allowed us to include documents from several sources.

Permission to include the Economic Declaration of the Fourth Conference of the Heads of State of Non-Aligned Countries was arranged through M A Chitour of the Algerian Embassy in London, and the Embassy also provided us with a copy of the speech of President Houari Boumedienne of Algeria to the First Conference of Sovereigns and Heads of State of OPEC Member Countries. The article "First, Second, Third and Fourth Worlds" by Barbara Ward is reprinted by permission of The Economist Newspaper Limited.

Permission to include extracts of the White Paper on "World Economic Interdependence and Trade in Commodities" was arranged by Janet Hewlett-Davies of the Prime Minister's Office in London and Mrs B J Jones of Her Majesty's Stationery Office. Mrs D Corsbie of the Guyana High Commission in London kindly arranged for us to have permission to reprint the statement of the Prime Minister of Guyana, Mr Forbes Burnham to the Conference of Commonwealth Heads of State in Jamaica in May 1975.

Mr Jørgen Milwertz, Head of the UN Centre for Economic and Social Information in Geneva gave permission for us to reprint the Declaration and Programme of Action on the New International Economic Order from the Sixth Special Session of the UN General Assembly held in New York in April 1974.

Mr Bernard Chidzero, Head of the Commodities Division of UNCTAD provided us with a number of documents from UNCTAD which were relevant to the theme of this book, and permission to include the UNCTAD document "An Integrated Programme for Commodities" was given by Mr N Groby, Head of the UNCTAD Office of Administration.

We would like to thank Mr Roy Baker of Plenum Press for his help and encouragement and finally we would like to thank Mrs Andrea C Brown for all the work involved in typesetting the manuscript of the book.

October 1975   Paul Rogers
                                                        Anthony Vann

CONTENTS

Part I

RESOURCES, DEVELOPMENT AND PRODUCER POWER

THE GLOBAL ECOSYSTEM AND HUMAN ACTIVITY

Anthony Vann

Introduction

Over the last three hundred years or so, man has concentrated his scientific efforts on discovering the 'key' by which organisms function and survive. Central to most of these studies has been the organism as an individual and indeed many of the more remarkable findings have been at the cellular and subcellular levels. Undoubtedly this reductionist (Commoner, 1971) approach has led to significant advances in the applied biological sciences, but at the same time it has reduced the amount of attention given to the study of the organisation of living systems at the collective 'macro' level. While both approaches relate to individual and collective survival, the emphasis is rather different in each. While 'systems' have been studied empirically over a long period, it may perhaps seem surprising that fundamentally important aspects of environmental science were apparently undeserving of rigorous attention for so long. However, this must be seen in the context of the immediate social environment of the major centres of the scientific community. Catastrophies tended to be concentrated within the military and political fields. Threats to the survival of man came from disease or thermonuclear warfare.

However, two significant episodes in the last fifteen years or so brought environmental matters to the fore. Environmental pollution caused by the waste products of human communities, particularly in the industrialised countries, was rediscovered during the 1960's. That it could be a hazard to health had been shown long before, and many accounts of working and living conditions in the nineteenth century can be found in the writings of contemporary social reformers. In response to the threat of that pollution and the accompanying and related resource depletion and supply problems of the 1970s, the emphasis in education and research work in the Natural Sciences has been redirected.

The relationships between organisms, and between organisms and their environment were synthesised in the 1930's into the 'ecosystem concept'. The ecosystem is an organised assemblage of plants and animals with their environment. As a concept, it integrates much of the confusing detail of natural communities (eg. diversity) in the pursuit of functionalism. It can also apply at any level or scale, since it allows for import and export to and from the basic model. Some levels are clearly more suitable for study than others, the floristic and faunistic compositions and the time available for study being important determinants.

In the context of man's activities, it is the global level which has been commonly chosen. As many observers have commented, it may well be the holistic view of earth from space which will prove to be one of the most valuable conceptual aspects of the development of space technology.

While ecosystems may be examined in a number of ways, employing a wide range of currencies, there are two principle components which receive most attention, for obvious functional reasons, namely energy and minerals.

Energy in Ecosystems

Energy flows through ecosystems, and is changed from a non-random to a random state in the process. Random energy in the form of heat is lost at each transfer step within the ecosystem — for instance, when one organism consumes another. It is also lost when organic material not associated with living organisms is consumed by decomposer organisms, and at all points within ecosystems due to the energy requirements of continued respiration. Nutrients, on the other hand, while being accumulated from the environment and 'fixed' into living material mostly at the primary producer level and partially degraded again to simple materials at the various consumer levels, tend to be retained within stable ecosystems. The overall process of ecosystem metabolism is therefore one of diffuse energy (sunlight) being fixed in chemical bonds at the producer level and progressively degraded into random energy by transfer between, and maintenance costs within a number of consuming stages, with due consideration being given to imported and exported materials at all stages.

The overall balance of the energy flows of the Biosphere was summarised by Reynolds (1974) in the equation:

$$\frac{\text{Solar}}{\text{Energy}} + \frac{\text{Tidal}}{\text{Energy}} + \frac{\text{Geological}}{\text{Energy}} + \frac{\text{Fuel}}{\text{Energy}} = \frac{\text{Radiated}}{\text{Energy}} + \frac{\text{Stored}}{\text{Energy}}$$

Reynolds gives figures of $173,000 \times 10^{12}$ watts per year input of solar radiation power input, 3×10^{12} watts per year input of power through tidal activity, 32×10^{12} watts per year flow from the earth's core itself and 6.3×10^{12} watts (1970) per year derived from the consumption of fuels. The average change in the energy storage of the Biosphere is small so that energy radiated each year roughly balances the inputs on the left hand side of the equation (the largest of which is clearly solar energy). As Reynolds further points out, any transformations of energy within the biosphere will not affect the energy balance of the biosphere provided that none of the items of the overall equation change in magnitude.

Of the energy reaching the earth's surface most is reflected as short wave radiation (about 30%) or converted to heat after absorption in the biosphere (about 47%). A further 23% provides energy for the processes of the hydrological cycle (evaporation, etc). This energy is ultimately lost as long-wave radiation, although a little is stored within the cycle (Figures of King Hubbert, 1971). Estimates of energy fixation by green plants have shown that only a small proportion of the total available energy is utilised. Loomis and Williams (1963) showed that 5.3% of total incident energy (or 12% of visible light) could be stored in the products of photosynthesis. Total daily insolation was taken as 500 calories per square centimetre. These workers showed that this potential efficiency is rarely approached by crops in the field. Temperature, water and nutrients can and do influence productivity. Transeau (1926) for example showed that net production in a cornfield was 1.2% of total energy available.

Net production provides the initial substrate for the chain of consumers in the ecosystem. The efficiency of transfer of energy along the chain of consumers (herbivores, carnivores, top carnivores) is about 10 per cent (weight gain as a percentage of intake), although values have been recorded greatly in excess and also much lower than this guideline figure. The substantial losses at each transfer of energy are due to such processes as ingestion, losses in faeces etc., while a continual supply of energy is needed for respiration at all stages. Further losses occur due to some of the material of the previous trophic level being unavailable to the consumer. In agricultural systems, only part of net production is available as harvestable and available crops; the remainder, comprised of roots, leaves, stems and so on, will form the basic material for another food chain, which itself may include a number of trophic levels. Man is clearly just one component of such complex food webs, and is therefore in competition with a number of other life-forms (of which insects and fungi are the most notorious) for the available net production.

The process of development of human communities has led to a substantial change in the structure and productivity of ecosystems. While ecosystems have become more and more simplified, and the yield of harvestable material was increased, the amount of energy subsidy required to maintain these simplified systems and their high organic productivity has increased greatly. Newbould (1972) summarised the self-sufficiency ratio of a number of countries and cultures. He showed that the ratios of food energy to extraneous energy inputs were 1:0 for Kung Bushmen, 1:1 for subsistence agriculture, 1:20 for Britain, 1:32 for the USA, the World average being 1:12. Most of the input of energy in the 'advanced' agricultures is derived from fossil fuels, and in Britain accounts for 3.9% of total primary fuel consumption (Blaxter, 1975). Man is therefore becoming increasingly dependent on a food production system which for its maintenance and productivity requires a substantial input of fossil fuel energy. Agricultural systems are rapidly becoming black boxes in which fossil fuel energy is turned into food energy. This process is unfortunately not apparent to the casual observer. An equally undesirable aspect of this process is its export to less developed regions of the world, many of which have neither substantial reserves of traditional energy resources nor substantial foreign exchange surpluses with which to purchase them. Their dependence therefore on 'energy exporters' must to some degree increase.

It is not, of course, simple calorific value with which we are most concerned, but also quality and nutritional value. There therefore has to be some sacrifice of overall calorific output for the production of a diversity of products (Odum 1971). To a large extent this has been done by exploiting the primary consumer level of the food chain. The efficiency of calorific value transfer from plant material to animal may be increased by the use of extraneous energy for keeping the animal warm, saving food energy normally expended in walking, and so on. This factor contributes significantly to the overall energy subsidy of the more 'advanced agricultural' systems. That is not to say that the practice of keeping animals indoors and bringing food to the animal is a recent development. It is likely that this was a common system of animal husbandry in the late Atlantic period of British post-glacial history. One important feature, however, of the energy subsidies currently operating in advanced agricultural societies is the release of land for cultivation which was previously used to support draft animals, whose activities are now performed by agricultural machinery. Such trends towards energy subsidy of food production may be of relatively little significance (if unnecessarily profligate) in countries where subsistence agriculture would still be capable of supporting the population in its nutritional needs. However, where the maintenance of the population is currently dependent on the continued supply of non-renewable resources to agriculture (and principally energy in direct or derivative form) in order to maintain an energy-subsidised high productivity, the long term prospects for that agricultural industry are unpredictable. However, long before energy supplies themselves are limiting the price of these supplies is likely to increase the costs of energy subsidised agriculture dramatically. Energy consumption in agriculture in Britain accounts for only a very small proportion of the total extraneous energy budget, though it is important to take account of the further energy subsidy contained in food and animal feedstuffs imported in quantity from other parts of the world and which are equally vulnerable to energy supply and price fluctuations.

Hence, it is not primarily the consumption of energy in producing food which is likely to impose limitations on agricultural productivity, but the nine-tenths of extraneous energy (in the case of Britain) which is used elsewhere. From a ecological point of view, however, the use of energy in agriculture is of the greatest significance.

In 1972, western industrialised countries consumed 34 per cent of their annual extraneous energy budget in the industrial sector, 24 per cent in the domestic and commercial sector, 20 per cent in the transport sector, and 22 per cent in electricity generation (figures derived from Connelly and Perlman, 1975). The growth of each of these sectors has not been constant, but has been related to the pattern of development of the economies of the Western industrialised nations. For instance, energy consumption overall has been increasing at a rate of about 5 per cent each year in the United States in recent times, although the contribution of the electricity sector by contrast has been growing at a rather greater rate, around 9 per cent (figures of Cook, 1971).

Traditionally, this global ecosystem energy supplement has been supplied by the three major fossil fuels, coal, oil and natural gas. Up to and during the early part of the nineteenth century wood and similar vegetal material, together with water power in limited quantities, had provided the major energy supplement. But later in the nineteenth century, coal became the dominant fuel in the industrialised nations. The use of coal has increased annually at a rate of 3—5 per cent since 1860, except for the period between 1914 and 1944 (King Hubbert 1969). Estimates of coal reserves have been remarkably consistent over many years, those of Averitt (as described by King Hubbert, 1969) showing that only a very small proportion of the available coal reserves have been used to date, and that the bulk of those reserves are likely to be consumed between the years 2040 and 2380. Substantial reserves lie in parts of Asia, the USSR and North America. Smaller but highly significant amounts occur in Western Europe. However, Africa, Australasia and South America possess relatively small reserves (King Hubbert 1971). The concentration of the resources in restricted areas of these continents further exacerbates their irregular global distribution.

The use of oil, relatively unimportant until the late nineteenth century, increased at a rate of almost 7 per cent each year from 1890. Ryman's estimate (see King Hubbert, 1969) inferred that it would take 64 years for the bulk (80 per cent) of world crude oil reserves to be consumed. The distribution pattern of oil fields is well known, and the consequences of that distribution are becoming clear; the oil fields discovered in recent years in the North Sea, Alaska and so on remain fairly minor in comparison to the reserves within the OPEC group of countries, though undoubtedly very important within their local and regional contexts. There will be, however, a probable extension of oil reserves by the utilisation of oil shales and tar sands. Surrey and Bromley (1973) have shown how possible recoverable oil from shales could increase total crude oil reserves between three- and six-fold, with further substantial additions from tar sands and, in particular, coal conversion. Furthermore, they stress that such increases would involve only relatively moderate price rises and would involve the employment of current technology, suitably developed. The environmental cost of such operations may, however, be quite high, at least in local terms.

Natural gas occurs in most of the oil fields of the world, yet in many areas has only recently achieved significance as a fuel resource. In Britain there has been a marked increase in the use of natural gas in recent years. It accounted for only 0.5 per cent of total British energy consumption in 1965 but had risen to 11.5 per cent by 1973. Much of this rise has, of course, been at the expense of gas previously generated from coal.

Energy derived from fissionable Uranium (the prime nuclear fuel) will undoubtedly assume greater significance in many parts of the world in the next few decades; Surrey and Bromley (1973) for instance show how cheap uranium should continue to be in adequate supply for the next thirty years provided that resource exploration continues at an adequate level. Ultimately, the development of further breeder reactors which would greatly extend this supply is of the greatest importance from a purely energy supply point of view. The problems evolving with the continued production of radioactive wastes may however prove to be both socially and environmentally unacceptable. Current public disquiet over the rapid expansion of the nuclear power industry could be expected to increase dramatically following any significant malfunctions or accidents. The use of thermonuclear energy derived from nuclear fusion under controlled conditions is still dependant on the appropriate high-technology developments. If harnessed it could be capable of providing very considerable amounts of extraneous energy at high efficiency into the far distant future. However, it is unlikely to have any significance for several decades yet.

The increased use of these sources of energy would clearly alter the amount of energy entering the biosphere from fuels in the equation of Reynolds already described. Fusion energy in particular may give man the opportunity to increase this biosphere energy input very considerably indeed. Already there is some concern that the energy balance of the earth is changing, despite the fact that fuel energy inputs are very small in comparison with solar energy inputs in particular. It would therefore seem unlikely that such changes are attributable to the release of fuel energies, though increased production of carbon dioxide and its release into the atmosphere may have an important effect on the energy storage factor of the biosphere. This effect — the retention, on balance, of more long-wave radiation — may be partially mitigated by the production of dust from a variety of human activities (principally industrial) and perhaps the disposition of ice in the upper atmosphere by aircraft. Recent climatic changes in the Sahel, Ethiopia and other parts of the world have been attributed to gross environmental modification by man, while Woodwell (1970) showed that the 10 per cent increase in atmospheric carbon dioxide content may have increased net primary productivity by between 5 and 10 per cent, increasing the stored energy of ecosystems by an appropriate amount in relation. That the carbon cycle of the biosphere is homeostatic to some extent is well known. The question must arise, however, as to how homeostatic it may be.

The use by man of renewable sources of energy may still have some minor environmental side effects, but in general their use simply modifies the flow of energy from solar, tidal or geological sources to radiant energy. Hydro-electric power deriving its energy from the hydrological cycle of evaporation and precipitation may be capable of substantial development in certain parts of the world with obvious geomorphic advantages. This has already occurred in some areas (eg. Scandinavia), while in others (eg. much of Africa) the potential is quite high though associated with a number of very significant side effects such as the spread of water-borne and associated diseases where HEP schemes are linked with irrigation schemes. Tidal power and wave power are currently little used, those countries with the appropriate coastline configurations and tides being the most advantaged with regard to the former. Few countries have yet developed tidal power to any great extent, probably because other energy-yielding resources such as the fossil fuels have been more readily available. The development of tidal power is, however, unlikely to have very great global significance, though it may prove very important in regional terms. Similarly geothermal power is immediately available in areas with certain types of volcanic activity, but may become a much more significant source of energy if deep drilling is used to tap heat energy at depth in the earth's crust. Estimates vary widely as to its possible importance over the next century.

Solar energy, as a diffuse energy source, provides the greatest energy input into the biosphere. A small proportion is trapped (q.v.) in the products of photosynthesis but the vast amount contributes only to the overall energy balance of the biosphere. There is considerable potential for the utilisation of solar energy, but mostly on a local scale since a large 'trapping' area is required for large solar power plants capable of supplying towns and industrial plant. That solar power seems to have escaped the plans of the majority concerned with building design and construction to date is remarkable, but it would seem likely to be deployed extensively in the near future. The conversion of solar energy to electricity is currently both capital expensive and inefficient but the use of solar heating is already feasible and effective. Such developments, and more substantial ones such as solar furnaces, only modify the flow of energy from solar input to radiation output within the biosphere, with a small element of storage, and unlike the traditional fossil fuel resources are not associated with gross environmental changes or traditional pollution problems. Even the limited use of solar energy, in domestic installations for instance, would release fossil fuel reserves, whose flexibility in use in the petrochemical sectors of industry makes them almost unique.

Per capita energy consumption in the world in general is in fact increasing at a more rapid rate than in the western industrial states alone. Starr (1971) estimates that it could be a further hundred years before the overall average per capita consumption reaches the level of the United States at present, and three hundred years before the two are equal. Starr also points out that if the standard of living of underdeveloped countries rose over the next twenty-five years to that of the United States at present there would be a ten fold increase in energy consumption over present levels. Since many underdeveloped countries aspire to western levels of affluence, there is clearly a will to approach this target. Overall consumption is likely to increase with population expansion, which itself may greatly depress expected per capita increase in consumption. Since so few underdeveloped countries possess substantial reserves of the traditional fossil fuel energy resources, the cost of imported energy will become an increasingly important survival factor for much of mankind. Some countries, however, may benefit from imported nuclear technology, others from hydroelectric power developments or direct solar power. In the latter case, such countries would not only retain a substantial degree of energy self-sufficiency but employ a non-polluting and renewable source of energy at the same time. The extent to which such concentrations of energy supply are necessary to the economies of countries which have substantial direct labour (human energy) resources is a matter of considerable ecological interest, since human energies can be largely supplied from renewable biotic resources, which in turn may or may not be energy subsidised.

Minerals in Ecosystems

In natural ecosystems, there is a well-known and characteristic cycling of the major nutrient minerals. Minerals may be taken up from the environment by plants and animals, incorporated into organic material or retained in a structural or metabolic role, and ultimately returned to the environment by excretion or death or transferred to another organism in another trophic level. In this manner, the major mineral elements are merely 'lent' to organisms for a relatively brief period of time. All mineral material is returned to the environment at some stage, many entering a geological cycle in which some of the transfer and storage phases are of enormous duration. The rate at which some of the major elements circulate within ecosystems is of great importance. The amount of storage in different phases of such cycles and the amounts transferring between phases are important characteristics of specific natural ecosystems. The processes involved in cycling are not well known for all minerals — in the case of the minor or trace elements, for instance, they occur so widely in the environment as impurities (the frequency of occurrence increasing in some cases with expanding industrial activity) or in such large quantities over large areas as to diminish the importance of cycling. In the case of the major plant nutrients, however, (nitrogen, phosphorus, etc) together with carbon and water, the rates of flow in almost all stages of the cycle influence the structure, composition and productivity of ecosystems. Furthermore, there is some interaction between cycles of different materials at certain points, and the factors which the cycles influence (above) may to some extent influence the cycles themselves. Such feedbacks are important, obviously, in the homeostatic control of ecosystem stability. Interruptions of flow or changes in storage at any stage, if significant, may disrupt or modify community structure or ecosystem dynamics.

Man as an omnivore in organising and controlling natural community structure and composition may have modified the rate of flow or amount of storage of certain minerals in certain parts of their cycles. For instance, the increase in arable and pastoral activity increased the frequency of occurrence of nitrogen-fixing leguminous plants. Biological fixation of elemental nitrogen may account for over half of the total fixation of nitrogen annually (Delwiche, 1970). Improvements in land drainage may have led to decreases in denitrification, the process by which soil nitrates are reduced to nitrous oxide or elemental nitrogen. Perhaps the most important modification of a natural cycle by man may be the industrial fixation of nitrogen: Delwiche estimated that the quantity of nitrogen fixed industrially was doubling every six years, and that if the amounts fixed by leguminous plants were added to this it would exceed natural nitrogen fixation by up to ten per cent. So many aspects of the operation of the nitrogen cycle at a global level appear to be as yet poorly understood that the question arises as to whether ecosystems, both natural and 'artificial' may be changed as a result of a general increase in circulating nitrogen. To date this has manifested itself on a localised scale as, for instance, algal blooms as a result of entrophication.

Carbon provides the major vehicle by which energy flows through ecosystems. However, the 'storage' phase of this cycle involving the atmosphere has been significantly augmented since the beginning of the industrial revolution by carbon released from fossil fuels (q.v.), themselves representing the degraded and stored organic remains of plant which fixed atmospheric carbon dioxide in the geological past, or the animals of higher trophic levels. Bolin (1970) estimated that the current consumption of fossil carbon is of such a magnitude as to increase the carbon dioxide concentration of the air by 2.3 parts per million each year. Others (eg. Deevey 1958) have suggested that a substantial proportion of the increased carbon dioxide content of the air has come from more recent (geologically speaking) carbon deposits such as bogs and other peaty deposits. If so, land management assumes an even greater importance than hitherto implied. The possible consequences of such continued increases in atmospheric carbon dioxide concentration have already been discussed (q.v.). The consequences for man of even minor climatic change and perhaps adjustments in sea level would be of the utmost gravity. Since two thirds of the potential enrichment of the atmosphere is absorbed by oceans or incorporated in increased primary productivity or biomass in the terrestrial environment, (see Bolin, 1970) the risk of such major changes may be much less than anticipated.

Many animals use the materials of their environment in a structural role. The greatest difference between ecosystems which include man as a consumer or an environmental factor, and those which do not, is the vast difference in the deployment of 'minor' nutrients. Elements such as copper, zinc and iron are important in certain nutritional roles whereas others such as lead, mercury and chromium have no known important nutritional role. Such materials are used widely for a variety of purposes in human society. Where the 'use' of these materials differs from their role in the natural environment is in the quantities deployed and the durability of some of their forms. Some are deployed in toxic form or degrade to toxic products. In general their use tends to be highly localised. However, all mineral materials still conform to geological processes such as weathering, deposition and so on. As a regenerative process this occurs far too slowly to produce a closed or balanced cycle. While sufficient reserves of the ores of these minerals are available, the limitations on their use will be determined by the cost of extraction and processing (which will be largely a cost in energy terms), their possible undesirable effects within the human ecosystem (eg. lead and mercury) and any deliberate manipulations of supply by price or volume. Over a period of time, reserves of these materials will be depleted although authorities differ as to the rates of depletion.

In the 'Limits to Growth' study, the MIT team demonstrated that not only was the static index of expected availability of many important resources very low (eg. lead 26 years, copper 36 years, zinc 23 years) but that assuming a five-fold increase in known reserves and projected exponential rates of utilisation (based on current figures) the index is still most unfavourable (eg. lead 64 years, copper 48 years, zinc 50 years). While such figures may be open to differing interpretations and qualifications they give a certain sense of scale to the problem of resource depletion, a scale which may appear adequate from an economic point of view, but is very narrow from an ecological point of view. In any case, much will depend on the term 'economically viable reserves , since many of the processes between extraction of a mineral resource and the final product (including transportation, processing etc) are highly dependent on a substantial energy subsidy. Since energy derived from all current major sources is becoming relatively more expensive, it is likely to be this cost which limits the use of many of the more energy-subsidised materials, rather than the absolute supply of the raw material. In this way, therefore, human ecosystems are likely to reflect increasingly natural ecosystems inasmuch as the form of ecosystem (and international) currency is becoming energy, or the resources from which it can be derived.

Four prime factors may progressively influence the availability of mineral resources. Firstly, the prospective input of materials derived from new terrestrial, marine and sea-bed sources will undoubtedly extend the continued availability of a number of key materials. Since sea-bed resources are rich in certain materials which are less common, or highly localised in distribution, in the terrestrial environment, there may possibly be a modification in technology to accommodate this. The exploitation of sea-bed resources is likely to be qualified by the political and economic outcome of the continuing conferences on the jurisdiction and exploitation of the oceans (the "Law of the Sea" conferences) and by the energy subsidy required in the extraction and processing of the raw materials gained.

Secondly, human ecosystems could more nearly parallel other ecosystems in cycling far more of the basic materials used by society. Many materials are already extensively recycled, not only metals (eg. copper 35%, aluminium 20%, steel 34% — Connelly and Perlman, 1975), but also some vegetal materials such as paper could achieve greater unit-use. Recycling, of course, accommodates both of the commonly stated criteria of conservation, those of resource protection and resource development. While recycling undoubtedly saves the use of energy in the extraction and processing of materials (which may be very substantial in the case of some metals), there is a further energy input to be set against this in the form of collecting, sorting and reprocessing. To put this in context however, even a higher degree of recycling could not supply the continuing and increasing demand for most of these materials.

Thirdly, the substitution of one material for another in short supply (for whatever reason) may both prolong the 'life' of the latter resource while shortening that of the former. Substitution developments have, in the past, frequently involved the use of more energy or the use of derivitives of fossil fuels as substitutes themselves. Much therefore has depended on the availability and cost of these resources. However, the possibility of substituting natural renewable (agricultural/silvicultural) products for synthetic ones — the reverse of the process which has occurred through much of the twentieth century — may be limited (again) by the energy cost of the production of the natural products and the alternative uses for which the appropriate land areas may be used — such as the growing of food crops.

Lastly, the possibility that industrialised man may choose to actively conserve certain key materials, or choose not to use them, cannot be ruled out. Presumably such materials would be in short supply, or would have been shown to have serious side-effects in production or use, and would have been the subject of a programme of public education. Already, small groups of people exercise voluntary abstinence of meat eating on certain days to attempt to extend world grain supplies; other groups use no synthetic pesticides in producing food. Such activities are however restricted to fairly small sectors of society at present.

The aspirations of both developed and less developed will undoubtedly give rise to dramatic rises in demand for a wide variety of materials derived from the Biosphere and the earth's crust. However, an over-riding factor which compounds the demand is, of course, the current rate of human population increase. Such a familiar topic requires relatively little in the way of statistics to emphasise its importance. The current rate of population increase stems from a high natality rate coupled with a low rate of mortality. The causes of these two factors have been the subject of extensive discussion elsewhere. The net effect, however, is a growth rate overall of about 1.8 per cent, representing a doubling time between 30 and 40 years. The overall rate is, moreover, made up of widely differing rates in different parts of the world, South America having a rate of population increase of about 3.0 per cent, while that of Europe is about 0.7 per cent (figures from McHale, 1967). There are few indications that the population of the world, currently about 4.0 billion, will fall short of the expected 6.5 billion by the year 2000 as given by United Nations statistics. Furthermore, we face the doubly alarming phenomenon that in many countries with a high proportion of their populations in the younger age groups, an immediate reduction in birth rate to 'replacement levels' would still result in a very substantial increase in population before stabilisation. The problems of providing sufficient food for the rapidly increasing population of the world are well illustrated by the fairly modest productivity achievements of the so-called 'green revolution', its dependence on inputs of fertilisers and energy, and its social side effects (see Dahlberg, 1974). The carrying capacity of some parts of the world may already be exceeded by the resident population, but taken as a global figure, the population of the world capable of being adequately supported nutritionally was estimated by Gates (1971) to be between 10 and 12 billion, allowing for improvements in overall productivity and protein production together with increased oceanic and terrestrial exploitation. Clearly a global population of such magnitude could be highly susceptible to short-falls in production and the energy supplies required to maintain production. Gates also showed that a global population of about 8 billion would retain greater flexibility in terms of risk avoidance. The MIT 'Limits to Growth' study also showed that the available arable land area would more or less balance the amount of land required at present productivity levels to maintain the world's population about the year 2000. Revisions of farming practice particularly in terms of the types of crops grown and the use of the primary producer level for protein supply rather than the herbivore level could constitute some of the most important measures — in other words, the growing of crop species of high protein content, vegetables and grain and root crops, at the expense of the grazing animal. The world food supply could be considerably extended by such practices, which would of course involve quite substantial dietary modifications particularly in the high meat-consuming areas of the developed world. Such measures, which could be difficult to implement from the point of view of public inertia, could only be seen as temporary expedients, however, in the light of the current rates of population increase.

Developed countries show a much higher energy and raw material consumption per capita than less-developed countries. If the latter aspire to similar levels of consumption, though perhaps in different consumption patterns, demand for resources will increase massively. Irrespective of whether the relationship between energy consumption and GNP per capita is linear or curvilinear, (energy consumption per capita levelling off at a level at or near the current level of the United States), it would seem unlikely that world production of energy yielding resources could supply a global level of demand equivalent to that of the US for very long, in ecological terms. (For a discussion see MIT 'Limits to Growth' study, and Science Policy Research Unit of Sussex University 'Thinking about the Future' critique). By illustration, Newbould (1974) suggested that a fifteen-fold increase in demand for resources could occur. Gone are the days when increasing populations could be seen as an increasing resource in terms of labour supply and a growing market for agricultural and industrial products. They represent increasing demands on already stretched resources.

While the projections of availability and utilisation given in this account are subject to wide interpretation and qualification, the trends in resource use and population growth, and the implications of this within the context of the global ecosystem, are fairly clear within an ecological time perspective. It is likely that the twin problems of population increase versus food supplies and the cost and supply of energy will dominate world affairs for at least several decades not only in an economic and humanitarian sense, but in a political and possibly military sense as well. After that time, a new world order may become established and stabilised. The fossil fuels will continue to dominate as major suppliers of energy for the rest of this century and much of the next but rationing by either price or volume and probably both can be expected in the near future. This is highly likely in the case of oil because of its high flexibility in use in the human ecosystem. Conflict over such resources could amount to little or escalate quickly in view of the need for conventional energy sources for armed conflict of a conventional nature. As competition for limited resources increases, it will doubtless be manifested in the classic variety of individual and collective behavioural responses. In the medium term, increases in fossil fuel use could have serious climatic consequences, but since the climate of the world has shown marked fluctuations in the recent past due to purely natural causes, it may be difficult to attribute any changes to anthropogenic factors. However, this should still remain the greatest argument against complacency and continued environmental modification and for further research on global ecosystem dynamics. The nuclear energy programme will probably accelerate in the short and medium term, particularly in the field of breeder reactors, until the first major plant malfunction, after which the programme will proceed much more cautiously or perhaps even halt. Solar power collectors and to a lesser extent wind power and geothermal power collectors are likely to be developed and deployed on a wide scale. The development of fusion power plants remains a great unknown and one which could potentially pose the greatest problems with regard to the balance of global energy flows.

Certain key minerals may soon run into short supply but the availability of substitution materials may be limited — agriculturally by the need to produce increasing quantities of food materials, and synthetically by the availability of fossil fuels. This may be one of the strongest arguments for the conservation of certain fossil fuels. Furthermore, it is hardly a good strategy to assume that technology will provide answers in time. New developments should be reviewed in the context of their global impact and place in man's ecosystem and without the urgency for new technology to be put into service immediately.

The population problem is already critical. Many parts of the world are already experiencing natural 'controls' of population size (as they have for so long) and it is certain that many more will do so in the very near future. It would appear that little can be done to avoid this. Flexibility and time are no longer available, which only emphasises the efforts and sacrifices which will have to be made if tragedy on an unprecidented scale is to be avoided. Since it is probable that renewable energy sources are likely to be dominant in the medium and long term, now is the time to use the resources at our disposal (including fossil fuel energy) to adapt society to a more ecologically relevant way of life. Now is the time to retain key raw materials by conservation and recycling policies, to invest effort and energy in improving agricultural systems and to revise modes of transport or to remove the need for such massive movements, not later when energy and resources may be scarcer. The advocates of market forces being alone allowed to control the flow of resources and their cycles of production should examine closely the choices which may face society as key resources approach scarcity. But for the present, the 'adaptability gap' in western society is still open.

To set against this, and for obvious and understandable reasons, there is a great deal of inertia in human society at all levels. Man has spent a great deal of effort in overcoming limitations on his population size and activities, and modifying ecosystems to further his own chances of survival, while creating further — perhaps more intractable — problems in the process. If we do not impose limitations on our own activities and population growth, nature will ultimately impose limits for us — limits which are likely to follow the classic evolutionary pattern of natural selection which human society has strived so long to circumvent.

Bibliography

Averitt, P. (1969) "Coal resources of the United States, 1 January 1967". Geological Survey Bulletin 1275. Government Printing Office, Washington DC.

Blaxter, K. L. (1975) "The Energetics of British Agriculture". *Biologist 22.* 14–18.

Bolin, B. (1970) "The Carbon Cycle". *Sci. Amer.* Sept. 1970.

Commoner, B. (1971) "The Closing Circle". Publ. in Britain, 1972. Jonathan Cape.

Connelly, P. and Perlman, R. (1975) "The Politics of Scarcity". Oxford Univ. Press and The Royal Institute of International affairs.

Cook, E. (1971) "The Flow of Energy in an Industrial Society". *Sci. Amer.* Sept. 1971.

Dahlberg, K. (1974) In "Human Ecology and World Development". Ed. Vann, A. R. and Rogers, P. F. Plenum Press.

Deevey, E. S. (1958) "Bogs". *Sci. Amer.* Oct. 1958.

Delwiche, C. C. (1970) "The Nitrogen Cycle". *Sci. Amer.* Sept. 1970.

Guyol, N. B. (1949) "Energy Resources of the World". Dept. of State publication 3428, US Government Printing Office, Washington DC.

King Hubbert, M. (1969) "Energy Resources" in "Resources and Man", A Study and Recommendations. Committee on Resources and Man. US National Academy of Sciences — National Research Council. W. H. Freeman.

King Hubbert, M. (1971) "The Energy resources of the Earth". *Sci. Amer.* Sept. 1971.

Loomis, W. F. and Williams, W. A. (1963) "Maximum crop productivity: one estimate". *Crop Sci. 3.* 67–72.

Meadows, D. H., Meadows, D. L., Randers, J. and Behrens, W. W. (1972) "The Limits to Growth". Earth Island.

McHale, J. (1967) "The Ecological Context: Energy and Materials Document 6. World Resources Inventory." Southern Illinois University.

Newbould, P. (1972) "The contribution of Ecology to the study of Human Ecology". In "The Education of Human Ecologists" Ed. Rogers, P. F. Charles Knight, London.

Newbould, P. (1974) In "Human Ecology and World Development". Ed. Vann, A. R. and Rogers, P. F. Plenum Press.

Odum, H. T. (1957) "Trophic structure and productivity of Silver Springs, Florida". *Ecol. Monogr. 27.* 55–112.

Odum, H. T. (1971) "Environment, Power and Society". Wiley.

Reynolds, W. C. (1974) "Energy: from Nature to Man". McGraw-Hill.

Starr, C. (1971) "Energy and Power". *Sci. Amer.* Sept. 1971.

Surrey, A. J. and Bromley, A. J. (1973) "Energy Resources". In "Thinking about the Future". Ed. Cole, H. S. D., Freeman, C., Jahoda, M. and Pavitt, K. L. R. Chatto and Windus and Sussex University Press.

Transeau, E. N. (1926) "The accumulation of energy by plants". *Ohio J. Sci. 26.* 1-10.

Woodwell, G. M. (1970) "The Energy cycle of the Biosphere". *Sci. Amer.* Sept. 1970.

OIL PRICES, OPEC AND THE POOR OIL CONSUMING COUNTRIES

Biplab Dasgupta

1. Introduction

In 1950, the year O.P.E.C. (Organisation of Petroleum Exporting Countries) was formed, the world oil industry was dominated by a group of seven oligopolistic major international oil companies, who were collectively known as 'majors'.[1] The latter owned 84% of world crude oil production, 74% of refining capacity and 70% of marketing activities[2] outside the United States and the Communist countries. They produced oil, researched on oil exploration, refined, marketed and transported oil; they knew everything about oil and did everything there was to do to run the industry. The seven majors also fixed the oil prices and decided on who was to sell oil, how much, and to whom; they were the intermediaries through whom countries producing and consuming oil transacted their business. The majors were the oil industry, no country or government in the less developed parts of the world, oil producers or consumers alike, could risk antagonising these vast corporate entities.

The last fifteen years of O.P.E.C. have completely transformed this picture. Whereas previously the majors fixed the oil prices, now it is the prerogative of the O.P.E.C. and its constituent oil producing countries to quote prices for their oil, whereas previously the majors controlled the distribution of oil, now individual oil consuming countries — both developed and underdeveloped — are eager to establish bilateral trade links directly with the oil producing countries. Whereas until recently the major oil companies owned the oil resources of the Middle East, by mid-1975 very little of that vast empire remains. Not that the large multinational major oil firms have disintegrated, but the whip is no longer in their hands.

One major objective of this paper is to trace the history of the growth of O.P.E.C. and its role in the rapid change in the correlation of forces in the world oil industry, and also to examine the factors which created the conditions favourable for the birth and the successful run of this mighty trade union of some of the richest countries of the world.[3]

A second objective is to examine the impact of the recent high oil prices, an outcome of the long sustained O.P.E.C. campaign for better rewards for its possession of a most important raw material, on the less developed oil consuming countries of the world. Have they gone broke? Is the high oil price the only or the main cause for their present misery? Could O.P.E.C. do anything to help the poorer countries? Would they? These are also some other questions to be reviewed by this paper.

We begin with brief accounts of other main actors in the oil drama besides the two groups of countries mentioned in the previous two paragraphs — the multinationals, the host countries of the multinationals the other developed oil consuming countries, and the Soviet Union (Section II). Section III examines the pricing system which existed before the sixties and the historical setting which led to the birth of O.P.E.C. In Section IV, we briefly examine the issues handled by O.P.E.C. at various times. Section V explores the political economy of oil; while in Section VI we examine the various consequences of high oil prices on less developed countries and the possibilities of wider collaboration between the latter and the O.P.E.C. countries.

2.   Multinationals, Rich Oil Consuming Nations and the USSR

In this section, we briefly introduce the three major components of the oil industry besides the O.P.E.C. countries and the less developed oil consuming countries.

2.1 Multinational Corporations

No discussion on the oil industry is complete without reference to its seven leading firms, collectively known as 'majors' or 'seven sisters'. Five of these are domiciles of the United States, one is British and one is Anglo-Dutch. The vast size and financial resources of the majors enable them to enter into costly, high risk, capital-intensive ventures like finding oil in Alaska or the North Sea. Their massive investment in research and development helps to maintain their technological superiority over rivals. Very few countries of the world possess the resources, the skilled manpower, and the intimate knowledge of the industry and the world market to bargain on equal terms with these mammoth entities.[4]

Three main features of these major oil firms are their (a) multinational character, (b) vertically integrated organisation, and (c) oligopolistic relationship with one another. Because these firms are multinationals, their activities often give rise to serious conflicts between their own *global* objectives and the *national* objectives of the host governments.

Whereas the oil companies would want to maximise the aggregate profit over *all* its operations in all the countries in which their affiliates and associates are functioning, the governments of oil producing countries are interested in increasing their own revenues from oil exports.

The vertically integrated nature of their operations, — that is their participation in all stages of activities in the oil industry, from exploration, development, and extraction of crude oil to transportation and marketing — has many interesting consequences. For example, a major consequence of it is that a great deal of commercial and trading transactions in the oil industry (which often involve many countries) are no more than *intra-company* transactions between affiliates of the same mother company. Hence, the prices charged and paid for are not the results of "arms length" deals; but are internal book-keeping prices involving various branches of the same corporate body.

The majors are close partners (as well as rivals) in an oligopolistic market which until recently was ruled by a commonly adopted price formula, detailed market-sharing arrangements, and various types of jointly owned and pooled operations. Caltex (jointly owned by two US majors), Standard Vacuum (jointly owned by two other US majors until the mid-sixties) Burmah-Shell (jointly owned by an associate of BP, and Royal Dutch Shell) are some examples of collaboration among the majors in the fields of marketing and refining. But more important examples until recently such as Aramco (Saudi-Arabia), Kuwait Oil Company, virtually monopolised the production of crude oil from the Middle East until recently such as Aramco (Saudi-Arabia), Kuwait Oil Company, Iran Iranian Consortium, or Iraq Petroleum Company. They are also bound with one another by several long-term crude oil purchase deals; as also by short term localised operations of swapping crude oil and products in order to minimise transport costs. (See Table 1).

The history of the oil industry is full of explicit, written, market-sharing agreements between the oligopolists. [5] Sometimes markets were divided up territorially (as in the 1905 agreement), some times by percentages (as in the 1928 'as is' agreement), and at other times by a combination of percentage shares and absolute amounts in a specific market (as in India during 1905—1928). Since the 1940s written agreements are unknown, because of the bar of US anti-trust legislations, but the behaviour of the majors in this respect is influenced by the unwritten code of conduct of the oil industry against encroachment into each other's territory. Breach of an agreement leads to 'price wars' from which none of the participants emerges unhurt. Until 1928 several price wars were fought among the majors; but since then the nature of price wars has changed. Most of the price wars of the last four decades have been fought between the majors collectively on one side, and the newcomers to the oil trade on the other.

Table 1   The ownership pattern in the operating companies of the major oil producing countries of the Middle East until 1973

Country	Operating Company	Standard Oil New Jersey	Socony Mobil Oil	Standard Oil California	Texas Oil	Gulf Oil	B.P.	Royal Dutch Shell	C.F.P.	Others
Iran	Iranian Consortium	7	7	7	7	7	40	14	6	5
Saudi Arabia	Aramco	30	10	30	30					
Iraq	Iraq Petroleum[1] Company	11.875	11.875				23.75	23.75	23.75	5
Kuwait	Kuwait Oil Company					50	50			

Notes:

[1] The ownership pattern of Iraq Petroleum Company was followed in the operating companies of Qatar, Bahrein and Abu Dhabi.

Alongside the seven major oil companies function a large number of smaller companies, known in the oil world as 'Independents' or 'minors'. C.F.P. a French company partly owned by the government, is the largest among them and is sometimes described as the 'eighth major'. Some of these so called 'minors' are quite large by almost any other standard excepting that of the oil industry, e.g. Philips, Occidental, Standard Oil of Indiana. During the sixties their relationship with the majors was one of hostility; the developments of recent years have brought these two groups of companies closer to each other.

2.2   Rich Oil Consuming Countries

Developed western oil-importing countries do not constitute a homogeneous group. For example, the United States, though a net importer of oil also happens to be the largest oil producing country of the world. Moreover, some of them (e.g. USA, UK and Netherlands) are to varying extents closely associated with oil-producing interests through the operations of multinational oil corporations which are domiciled there. Despite their being net importers, these countries earn much more foreign currency through profit-remittances of oil corporations than they spend for their very substantial oil imports. [6]

Furthermore, given the importance of oil both as a source of energy and as a strategic material for military purposes, these countries have traditionally strongly backed their own oil firms in the latter's conflicts with the host country governments and the oil firms of other countries; because through the major companies these countries retain access to the large oil reserves of the middle East. A good example of the involvement of this type was the international political crisis following the nationalisation of Iranian oil industry in 1951; when both the British and the US governments used their diplomatic influence over other countries to block successfully the exports of Iranian oil, which eventually led to the violent overthrow of the Mosadeq government in 1954. Conflicts between different mother countries for influence in the oil producing areas often led to the sort of compromise reflected in the ownership pattern of several oil consortia in the Middle East (See Table 1).

Other oil-importing non-Communist developed countries which are not associated with the operations of multinational oil firms are usually highly sensitive to higher prices of crude oil and their effects on balance of payments. Two of the largest oil importing countries of the world — Italy and Japan — belong to this group. Both of these countries have been keen in curbing the influence of majors by promoting local companies and encouraging exploratory work both home and abroad. Some other oil-importing countries, until the recent price increases, have been less enthusiastic about reducing the prices of imported crude oil, in the case of West Germany in view of its possible harmful consequences on its large coal industry and its very large community of coal miners.

2.3   USSR and China

Last, but not the least, is the USSR, the second largest oil producing country of the world and the only major exporter of oil outside the framework of O.P.E.C. [7] About four-fifths of the massive Soviet production is required for domestic consumption which is growing and would in future probably restrict Soviet exports. Half of the exports go to the East European countries and the other half to a large number of countries of the world, West Europe accounting for most of it. The most recent estimates of proved oil reserves show the Soviet Union ahead of any region outside Middle East and North Africa, and containing about double the proved resources in the United States. [8] The main obstacle to oil production in the Soviet Union is the vast size of the country, and the geographical spread of its oilfields. Many of its oilfields, particularly the new discoveries, are located in less accessible western Siberian parts where production is costly and time consuming, and the cost of transport is a major component of the ultimate consumer price in the developed western part of the country. [9]

Although not a major factor in the world oil market yet, China might emerge as a key exporter by the end of this decade. An oil importing country until recently, China's crude oil production jumped from 10 million tons in 1968 to 65 million tons in 1974, and it is expected to reach a high figure of 400 million tons by the end of the seventies. [10]

3.   World Oil Prices and the birth of O.P.E.C.

3.1   World Parity Prices

A major feature of the oligopolistic behaviour of the major oil companies is their adherence to a common world parity pricing system: a simple, easily understood, basing point formula. Until 1945 the US Gulf was the basing point; the price of oil anywhere in the world was fixed by adding transport costs from the US Gulf to that consumption point irrespective of where the oil actually came from. This formula was followed even when no oil was imported by a consumption centre from the United States; its primary function being to minimise disputes about prices among the oligopolists and the consequent risks of a price war. [11] One consequence of this policy, from the point of view of many oil consuming countries, was that they did not benefit from their possible proximity to oil producing areas, since for the purpose of price fixation it was only their distance from the US Gulf which mattered. For example, under this pricing system the price of oil produced in Burma was cheaper in London than in Calcutta because London was closer to the US Gulf.

In 1945, Persian Gulf was adopted as the second basing point for the purpose of pricing. This meant that for markets nearer the Persian Gulf the price of oil was equal to the f.o.b. US Gulf price plus the ocean freight from the Persian Gulf to that market, again no matter from where oil was actually imported. For a time, only the freight element was changed, while the f.o.b. element continued to be identical with the US Gulf price. But after 1948, even the f.o.b. prices in the two bases — US Gulf and Persian Gulf — drifted apart. By 1949, f.o.b. prices in the two bases changed so much that, at given ocean freights, the delivered prices of crude oil of the same quality from these two areas were equalised in the Eastern seaboard of the United States. This change — the decline of US Gulf as base and the rise of Persian Gulf as the base for calculating oil prices in Europe, Asia and Africa and a good part of South America — reflected the fact that the United States, which accounted for 70% of world exports in the 1930s, had, after the second world war, been reduced to the status of a net importer of oil.[12]

The term 'posted price' which came to use at that time referred to the price quoted by the major oil companies at which they were wishing to offer crude oil of a given quality from the Persian Gulf. Allowances were made for quality — a fixed rate of two cents per barrel per each degree A.P.I., a measure used for estimating gravity. Non-Persian Gulf sources fixed their f.o.b. prices by adding to the f.o.b. Persian Gulf price ocean freight from the Persian Gulf to that source. Ocean freights were standardised by using A.F.R.A. (Average Freight Rate Assessment), a scale determined by a panel of six leading tanker brokers of London set up by two major companies in 1949.[13] Given the gravity of the crude oil, the distance of its source from the Persian Gulf, and the f.o.b. price at Persion Gulf of crude oil of that quality, it was a matter of simple arithmetic to work out the price of oil at any particular or consuming point. Since every major company 'posted' the same price f.o.b. Persian Gulf, only one 'world parity pricing system' ruled the market.

These 'posted prices' were mostly notional, 'book-keeping' prices for transactions between crude producing and refining-affiliates of the same mother company. These were not prices determined by the interaction of forces of demand and supply in a free market. The manager of the refining affiliate was not in a position to look for other sources of crude; for him the demand curve for crude oil was price-inelastic, as was the supply curve of the crude-producing affiliate. For a vertically-integrated multinational company it was the difference between the ultimate price the consumer paid for oil products and the total costs of producing, transporting, and refining crude oil and the cost of marketing oil products which mattered. The prices at a particular stage of operation — crude production, refining etc. — were determined by a host of factors such as relative tax rates in different countries and on different operations, as also on a number of social and political factors. In fact there was a tendency to show a relatively high price for crude-producing activities and a low price for downstream activities, as these enabled the US-based companies to offset depletion allowances and taxes paid to oil producing country governments against their tax liabilities in the United States.[14]

3.2   The Soviet oil offensive and the rise of 'independents'

As long as the major oil companies remained in virtually exclusive control
of the production and trade in oil the 'Persian Gulf parity pricing system'
was impregnable. But this system faced its first serious challenge in 1959,
when the USSR — a country not within the sphere of influence of the
majors — emerged as a big exporter of oil. Until then Soviet exports were
irregular, and both the production and export of Soviet oil were frequently disrupted by riots, civil wars, revolutions, and world wars. By 1959,
Soviet exports became stable at about 20% of domestic production
(compared to 3% in 1950). In absolute figures the net exports increased
from a meagre 3.6 million tons in 1955 to 25.4 million tons in 1959.

There were several explanations for the rapid growth in Soviet exports
during 1959 and 1960. The main explanation was that Soviet oil was
cheaper, the price being between 10 and 25 per cent less than Persian
Gulf prices. Moreover, payments could be made in local currency or with
non-oil commodities. For example, Egypt paid with cotton, Cuba with
sugar, Greece with tobacco, Israel with citrus fruit, and Italy with steel
pipes. In fact, Soviet oil trade usually formed a part of a big package deal
involving crude oil, oil products, as well as other non-oil commodities,
and the prices of its individual components were not specified.[15] Furthermore, unlike the majors, the Soviet Union was prepared to help in the
construction of oil refineries, without asking for equity ownership. These
refineries, when constructed, were operating outside the control of the
large international firms, were not tied to any particular source of crude
oil, and hence were able to buy crude from non-major sources including
the USSR. The Soviet Union also emerged as an alternative source of
technology for oil exploration.

This sudden growth in Soviet exports immediately brought the latter
into open conflict with the established major international oil companies.
The low Soviet prices disrupted the standardised Persian Gulf parity
pricing system. Soviet supply made serious inroads into markets which
were traditionally supplied by the majors from the middle East. Even
many anti-communist governments bought Soviet oil to reduce their
dependence on majors. In Finland, the government forced the refineries
owned by the majors to refine Soviet oil; in India and Japan the majors
were forced to reduce crude oil prices to meet Soviet competition; in
Italy, Soviet oil was used by E.N.I. to successfully launch a campaign
against the majors to reduce oil product prices; in Ceylon the Government
nationalized marketing companies owned by majors for their refusal to
handle cheaper Soviet oil products. 'Only fools and affiliates pay posted
prices', became the joke in the oil industry; but soon it became difficult
even for the affiliates of major companies to stick to posted prices.

In her confrontation with the major oil companies, the USSR was not
alone. Her activities during the late fifties and sixties coincided with the

discovery of several new oil fields by a number of 'minor' companies which were not organizationally linked with majors; particularly in Libya, Algeria and Neutral Zone. Since they did not possess elaborate marketing facilities, these companies were willing to sell crude oil and products at lower than posted prices; and like the Soviet Union they were willing to build refineries in various countries if that assured them an outlet for crude oil for a long period. All these together helped to transform the conditions in the oil market from being one under virtually the control of a small group of large oligopolistic firms to a more open and competitive condition.

3.3 The birth of O.P.E.C.

It was against this background that O.P.E.C. was formed in 1960. The direct cause of its birth was the fall in Persian Gulf posted prices for crude from \$2.08 (for 34° Arabian Light crude) in 1958 to \$1.90 in 1959 and \$1.80 in 1960 due to competition from the USSR and minors. For the oil producing countries — who had been eligible since the early 1950s for a half share in the profits from oil — this decline in oil prices meant a lower revenue per barrel of oil sold; and it was primarily with the objective of resisting a further fall in posted prices that this organization was formed,[16] in a meeting of representatives from Iran, Iraq, Kuwait, Saudi Arabia and Venezuela in Baghdad in September 1960.[17]

O.P.E.C. was not the first attempt at collaboration among oil producing countries for evolving a common policy towards multinational oil firms.[18] In 1949 an official delegation from Venezuela visited Iran and a number of Arab countries; the advice the Saudi Arabian government received from the Venezuelans encouraged them to demand a half-share in oil profits of the companies (in place of about 22 cents per barrel paid as royalty until then), which became the norm in the Middle East after 1950. The Iraqi-Saudi agreement of 1953, for exchange of information and periodic consultations, was another example of such collaboration which often helped the Iraqis to negotiate better terms from the oil companies in line with their agreements with the Saudi Arabian government. During the fifties, Arab League (which was founded in 1945), became a forum for consultation among some of the major oil producing countries. In 1952, the Arab League sponsored a committee of oil experts for coordinating the oil policies of its members, and in 1954 a permanent Petroleum Bureau (renamed Department of Oil Affairs after 1959) was set up. In 1959, the first Arab oil congress was held in Cairo under the auspices of the Department of Oil Affairs of the Arab League. Many of the dissatisfactions of the oil producing countries — about oil, low prices, royalty being a part of the 50% share, control of the industry by foreign firms, non-development of downstream activities in oil producing countries etc. — were voiced time and again in these forums but without much effect.

It needed the decline in posted prices with the growth of Soviet oil exports, and the threat of a further decline in future, for the major oil exporting countries to appreciate the necessity of collective action. Paradoxically enough, the soviet oil exports, although competing with the Middle East oil, contributed greatly to an improvement in the bargaining position of the oil producing countries vis-a-vis the major multinational oil firms. Previously they were dependent on the latter for market technology and managerial expertise. To the extent that the Soviet intervention weakened the control of the majors on the markets of the oil consuming countries and made the latter more knowledgeable about the affairs of the oil industry, the problem of finding buyers of oil was no longer that acute. Whereas during 1951—54 the nationalised Iranian oil industry could not find buyers of oil because of a boycott by the major oil companies, such a situation could not arise after 1960. Moreover, while technologically the oil producing countries continued to be dependent on foreign sources, they could now hire personnel or technology from concerns not associated with the major companies. For example, when the French-owned C.F.P. withdrew technical personnel from the Algerian refinery they could be replaced by the East Europeans; and the Soviet help in developing the State-owned North Rumaila oilfield boosted the confidence of the Iraqi government in its dealings with majors.

It was not until 1963 that O.P.E.C.s' status as a collective body of oil producing countries was recognised by the oil companies. One of the first successes was to prevent any further decline in 'posted prices'. Not that oil was actually sold at 'posted prices' in the market by the companies, in most cases during the sixties the 'market prices' were in fact lower than the 'posted prices'. But for the purpose of calculating the tax-revenue of the oil governments (i.e. half-share of the profits) 'posted prices' continued to be the bases, while the differences between 'posted' and 'market prices' were paid out from the half share of the oil companies.[19]

4.   The role of O.P.E.C.

Over the fifteen years of its life O.P.E.C. has fought many battles with the major oil companies. The following is a brief summary of some of the major issues they raised in their confrontations with the majors.

4.1   Prices

We have already noted the success achieved by O.P.E.C. in stabilising 'posted prices' during the sixties. This was a remarkable achievement at a time when the supply of oil was plentiful thanks to new discoveries of oil deposits in Algeria, Abu Dhabi, Nigeria and Libya; and the competition was keen with the arrival of the USSR and independents in the market. The actual 'market price' was on average 35 to 50 cents less than the 'posted price' of $1.80 (for 34° Arabian Light). Without O.P.E.C. the oil companies could not be pressurised into absorbing this difference between the two sets of prices.[20]

However, the O.P.E.C. countries were not content with the existing posted prices, they wanted more. Their arguments were as follows:

(a) The existing crude oil prices were a small proportion of the price the ultimate consumers paid for oil products (like petrol, kerosene, diesel, or lubricants). Most of the difference between the ultimate consumer price and the f.o.b. price of crude oil accrued to the major oil companies (as profit and charges for refining, marketing and transporting operations) and the governments of the oil consuming countries (by way of taxes).

(b) Given the inflationary trends in developed Western countries, and the consequent increases in the costs of industrial goods which are imported by the oil producing countries, a 'stable price' for oil in effect implied a declining 'real price' over time. It was pointed out that, unlike most other commodities in the world trade, both the absolute and the real price of oil had sharply declined over the previous 15—20 years. For example, in 1948 the price of Persian Gulf crude was $2.18 per barrel (for 34° A.P.I.), compared to a price of $1.80 during the sixties. There was a case for halting this decline in 'real prices.'

(c) The O.P.E.C. countries argued that the main beneficiaries of low crude oil prices were the developed consuming countries of West Europe, North America and Japan who accounted for the overwhelming part of world oil consumption (about 69% in 1972, compared to 3% consumed by the Middle Eastern countries which accounted for 37% of crude production).[21] The low oil price encouraged the wasteful use of this precious source of energy (e.g. in the form of heavy petrol-consuming private cars), while it benefitted the oil producing countries themselves very little. A higher oil price would discourage wasteful uses, and benefit the oil producing countries by way of higher revenue.

These three arguments for higher price — for a higher relative share in the ultimate consumer price, for better terms of trade, and for avoiding wasteful uses — made little impression on the oil companies who were then struggling to sell oil at the given posted prices. Realised prices were low and declining for most of the sixties. Some experts — notably Maurice Adelman of MIT — confidently forecast a price of one dollar per barrel by the early 1970s. Adelman's argument was simple: the production cost of oil in the Middle East (at 8 to 20 cents per barrel) was too low in relation to the market price of $1.30 to $1.60, and he expected further competition among producing countries and companies to bring down the latter to a level nearer to the 'floor price'.[22]

Under this situation the only way the oil prices could be raised was by the collective decision of all the oil producing interests to control production and restrain mutual competition. During the sixties several attempts were made by O.P.E.C. countries to work out rules for 'pro-rationing' which failed to work.[23] While the large, established oil producing countries wanted to adopt the existing production shares; the new oil producing countries like Libya or Nigeria, whose current production was a small proportion of the ultimate potential for oil production were naturally opposed to such formulae for distributing the target O.P.E.C. production among its members. Secondly, no such arrangement could work without the co-operation of the USSR, the chief rival of the Middle East in the oil market, but for political and ideological reasons several oil producing countries were unwilling to collaborate with the former. Thirdly, there were serious political divisions between the so called 'progressive' (Iraq, Syria, or Egypt) and 'reactionary' (Saudi Arabia, Kuwait and Arab Emirates) Arab countries; and between the Iranians and Arabs, which made co-operation in the field of oil industry difficult. Last, but not the least, as long as the production of oil was in the hands of the majors, no such programme could work without their support. But the majors were opposed to production control by government, as it curtailed their freedom of action.

Conditions in the world market began changing in favour of the oil producing countries after 1970. Between February 1971 and June 1973 there took place a series of increases in oil prices — $2.18 in 1971 (February), $2.48 in 1972 (January), $2.59 in 1973 (January), and $2.90 in 1973 (June) — based on negotiations between the O.P.E.C. and the oil companies. This largely reflected (a) a substantial increase in the demand for oil (particularly in the United States) due to the low prices in the sixties, (b) the inability of the rate of discovery of new oilfields to match this increase in demand, and (c) the consequent fall in the reserve production ratio which created the fear of exhaustion of oil resources within 30—35 years. The 'market prices' for crude oil (34° A.P.I. Arabian Light) jumped from $1.30 in 1970 (January) to $2.70 in 1973 (June), and the gap between 'market' and 'posted prices' narrowed to a mere 20 cents per barrel at this higher level of prices; factors which explain the acquiescence of the major oil companies to higher posted prices. Secondly, whereas the Soviet oil was a major competitor to Middle East oil during the sixties, its exports to non-Communist countries were stabilised at about 50 million tons during the early seventies. The demand for oil was growing so rapidly within the USSR that she was unwilling to expand her export activities, in fact, she imported 8 million tons of oil from the Middle East for supplies to border areas which were less easily accessible from her own oilfields.[24] Moreover, unlike the early sixties when she was trying hard to capture new markets, the Soviet Union was now an established trader in many markets of the world and hence she felt no desire to engage in price competition. For broadly similar reasons, the independents also accepted the new higher prices.

The trend towards higher oil prices thus predates the 1973 Arab-Israeli war; but the latter provided the Arab oil producing countries with the political motive (and power, arising from their unity) for further and steeper increases in prices — to $5.12 in October 1973, and then to $11.65 in January, 1974 — a more than six-fold increase in four years. Such an increase was made easier by the concerted action of the oil producing countries to cut back production of oil by 5% per month for several months. In the panic which followed, some oil consignments were sold at even higher prices — $18 or more — but eventually the prices settled down to a figure around $11 per barrel. A major feature of this price increase is the fact that prices were quoted unilaterally by the governments of oil producing countries, and the companies were left with no option but to accept these.

4.2 Profit-sharing

The price of crude oil was however only one, though the major, element in determining the percentage share as well as the absolute revenue earned by the oil country governments from oil. Negotiations were conducted with the oil companies during the sixties on several other issues. A major issue was the question of 'royalty'. As we have already noted, until 1950 the oil companies paid the host governments 22 cents per barrel as royalty but no share in the profit; the profit sharing arrangement which was subsequently evolved allowed for the splitting of the profit equally between the host government and the company. But the host governments' 50% included a royalty component of 12½%. O.P.E.C. demanded that the royalty (reward to the 'landlord') should be 'expensed' from taxes, that is paid in addition to a 50% share in profit. It was not until the late sixties that this demand was met. In June 1974, the O.P.E.C. countries unilaterally decided to further increase royalty to 14.5%.

Another issue was related to the items of cost deducted from total resource by the companies before arriving at the profit figure. There was suspicion in O.P.E.C. quarters that the cost figures were inflated by the oil companies; and they demanded a closer scrutiny of these figures. By 1964, a ceiling of half a cent per barrel was put on one of the components of these expenses — the marketing allowances.

A third issue concerned the percentage share of profit accruing to the host governments. The O.P.E.C. countries resented that the question of fixing rates of taxation on profits of a company operating within their territories' jurisdiction should be a matter for negotiation. (Among the O.P.E.C. countries only the Venezuelan government reserved its rights to unilaterally fix such rates). There was also the feeling that 50% rate was not adequate. As a result of a series of negotiations the rate was increased to 55% in 1971.

4.3 Participation

Almost from its inception, a major objective of O.P.E.C. was to replace the oligopolistic control of international firms with its own control. First, an attempt was made to end the monopolistic control over production by the oil companies. Until late fifties in most of the oil producing countries of the Middle East only one foreign-owned company (e.g. Aramco in Saudi Arabia or the Iranian Consortium in Iran) enjoyed the exclusive right to explore and produce; but only a small fraction of the total concession area was actually explored by them. O.P.E.C. demanded the return to the government of the right of exploration of those concession areas which were not explored within a given time period. In the case of Iraq incidentally, a law had been promulgated in the early sixties which took away 4% of the area under concession from the oil companies, which led to a prolonged conflict between the latter and the government.

Secondly, the concession areas thus recovered were turned over to state-owned oil companies for exploration. The state-owned companies were viewed as powerful instruments of the government policy of acquiring technical and managerial knowhow about the industry with the long term objective of wresting the control of the entire industry away from the major oil companies. The technological help came mostly from the independents and the East European countries. In many agreements provisions were made under which the foreign partner bore the risk of exploration, but if oil was found a jointly owned organisation with the state-owned company was formed.

Thirdly, demands were made for participation of local companies in the equity ownership of the large consortia controlled by the majors (like Aramco or the Iranian Consortia). But it took more than a decade before these demands were met. In 1971, the Algerian government took over the majority control in the oil industry; and in 1972, the operating companies in Kuwait, Saudi Arabia, Abu Dhabi and Qatar agreed to hand over a 35% share to the government, and promised to transfer 51% of the equity by 1981. In 1973, the Iranian government took over the consortium, and after the war of October 1973, in January 1974, Kuwait took over 60% of ownership of K.O.C., followed within one month by the takeover of three U.S. based companies by the government of Libya. By the end of 1974, the government of Saudi Arabia decided to take over the full ownership of Aramco.

By early 1975, very little of the Middle Eastern crude producing empire of the majors remained in their hands.

4.4 Other issues

Some of the other issues raised by O.P.E.C. could be briefly listed as follows:

(i) the need for a joint policy for conservation of oil resources;
(ii) the need for undertaking the refining of crude oil within the oil

producing countries, as opposed to the world-wide trend of locating refineries near the centres of consumption;

(iii) the need for greater use of oil within the oil producing regions and for linking the policy of oil with the general policy of economic and social development.

5.   The Future of O.P.E.C.

By now, O.P.E.C. has come to be recognized as a formidable force in the arena of international politics. In 1974, alone, O.P.E.C. earned an extra revenue of 65 billion dollars over the 1973 figure; a figure which is only slightly less than the tax figure (72 billion dollars) of total export earnings of all less developed countries (including O.P.E.C. countries) in 1972.[25] Even O.E.C.D., the powerful group of Western developed countries, is running a substantial balance of trade deficit vis-a-vis O.P.E.C. Attempts by the United States to form a cartel of rich oil consuming countries to confront O.P.E.C. and pressurise it to reduce prices have failed. Most countries in Europe have preferred bilateral oil trade links with individual O.P.E.C. countries to multi-lateral economic or political sanction against it.[26] Some of them, especially West Germany, have been keen and eager to persuade the Arabs to invest in steel and other large enterprises.

The crucial questions to ask now are: (a) are the oil producing countries capable of running their own industry without the assistance of the majors? and (b) will they be able to sustain the high oil prices for long?

5.1   O.P.E.C. and the Multinationals

To answer (a), let us first consider the various roles played by the majors in the O.P.E.C. countries. As far as their 'risk-bearing' role is concerned (i.e. in investing in high risk and expensive ventures) the element of 'risk' involved is not very high in the Middle East, and in some of these countries (Saudi Arabia in particular) the production-reserve ratio is too low to encourage a massive programme for exploration; in Venezuela or Indonesia the situation is different, but whether the companies would be keen to invest there now is another question. At this moment, there is no shortage of funds for such activities in these countries, should they decide to go for exploratory drilling.

The O.P.E.C. countries are still dependent on foreign support in the fields of (a) technology, (b) management, and (c) marketing. For a long time the major oil companies ignored the local demand for the training of indigenous personnel; the presence of a handful of local personnel in high technical or administrative positions was hardly more than symbolic. It was only during the 1960s that through the state-owned companies and with the help of international "minors" some progress was made towards the training of local personnel. Even now, although the local personnel are capable of performing the day-to-day routine activities, they lack the confidence and skill to handle breakdowns. The oil producing countries differ between themselves in their ability to run the industry — from Iran, which has a surplus of technical personnel, to Abu Dhabi, which is almost totally dependent on foreign personnel for high-skilled jobs. With further training of local personnel and a greater freedom of movement of skilled personnel between the oil-producing countries (which is inhibited by political factors, such as the mutual suspicion between Iran and her Arab neighbours), this technological constraint would, one hopes, be removed in future. But even in the short run the technological dependence on the "majors" is not unavoidable. Unlike the situation 15 years ago, today there are many suppliers of technology in the world market, including the "minors", the East Europeans, the Japanese and the companies like ENI, and equity ownership is no longer a price which must be paid to foreign interests for buying technology.

Perhaps more serious than the technological constraint from a long-term point of view is the marketing constraint. Until very recently an overwhelming proportion of the total crude and refinery production passed through the hands of the "majors", who acted as intermediaries between the oil exporting and the oil importing countries. Because they owned many different types of crude, operated refineries with a wide range of product-mix, conducted their business in hundreds of markets with widely varying demand patterns for oil products and owned tanker fleets to carry oil from a large number of producing areas, it was possible for them to balance the demand and supply for different types of crude and products at the world level, if not singly at least collectively through long-term supply arrangements and temporary swapping arrangements. In contrast, the governments of the oil producing countries in the world do not own elaborate marketing networks, and because they rely on a limited range of crudes and products, it is not possible for them to satisfy all types of consumers or to meet short-term shortages on surpluses in individual markets. At this moment — with the cloud of oil shortage hanging over the industry — the marketing problem is not so serious; in fact, the buyers themselves are eager to sign up long-term agreements or to bid enthusiastically for crudes of all varieties which are offered for sale. But this condition may not last indefinitely; with further discoveries of oil and the possibilities of technological breakthroughs in producing non-conventional oil and oil substitutes, in future it would be necessary for the oil producing countries to build their own tanker fleets and to strengthen their own marketing arrangements. It is also necessary, for various reasons, for such marketing activities to be co-ordinated through an agency like O.P.E.C.; firstly, because the marketing sector is the stronghold of the large international "majors" and no member country of O.P.E.C. would be able to match their strength singlehanded; secondly, because such co-ordination would make crude supply more flexible from the viewpoint of consumers; and thirdly, because without this co-ordination there is the risk of the member countries following widely divergent pricing policies which are in line with their own individual development programmes.

How are the major international oil companies themselves viewing the current situation? The present situation is not hurting them financially; in fact the 1974 figures for most of them show high profits. With high prices, and their continuation of market activities, they should continue showing profits for years to come — what is more alarming from their long term viewpoint is their decreasing share of crude oil production and loss of control over the market. They are responding to the present state of uncertainty in the Middle East and the oil market by shifting away from there to 'politically safe' areas like Alaska or the North Sea; and by gradually moving out of oil and extending their interests to non-oil and non-conventional oil industries, like nuclear energy, shale rock and tar sand.

5.2   O.P.E.C. and the U.S.S.R.

O.P.E.C.'s ability to maintain high prices for oil would very much depend in future on the oil export policy of the U.S.S.R., the only major exporter of oil outside the former's framework. The relationship between the Soviet Union and O.P.E.C. contains both elements of complementarity and conflict.

The conflict arises from the role of Soviet Oil as an alternative to oil from the Middle East. Gains made by the Soviet Union during the 'sixties had almost always been at the cost of Middle East oil deliveries. On the other hand, their economic interests converge in their policies towards the major international oil companies. As we have already noted, the Soviet competition, by weakening the majors, also improved the bargaining position of the O.P.E.C. countries *vis-à-vis* the latter. Over time, the intensity of Soviet competition is declining; after having established a certain share in the world oil market, the former has now developed a vested interest in stable and high oil prices. Moreover, with the growing pace of industrialisation, the quantum of exportable oil surplus is declining in the U.S.S.R.

5.3   Non-Oil Substitutes and Contradictions within O.P.E.C.

In the short run, at least, the prosperity of the O.P.E.C. countries is assured. There are few non-oil substitutes for oil products, and there would not be much competition from other suppliers of oil. Even at the current high level of prices, oil is competitive with other energy sources. The demand for oil has continued to grow, even though it is growing at a lower rate than the pre-1973 rate.

What is going to happen in the future is anybody's guess. The high oil prices have certainly made investment on further research for producing oil at a lower price from tar sand, coal and shale financially worthwhile.

The possibility of meeting the bulk of the world's energy needs from nuclear sources (or from solar or hydro energy) also remains. Much will depend on the extent of cost-reduction effected by these research activities on alternative energy sources.

What is clear, however, is that the O.P.E.C. countries cannot afford to rely indefinitely on oil as the main resource of government revenue and export earnings. A time is bound to come in the future when the 'oil bubble' will burst, and it is in their interest for the O.P.E.C. countries to invest oil revenue in other industries and activities, and thereby to reduce their dependence on oil before it is too late. O.P.E.C. countries are conscious of this need, which explains the high priority given to the transfer of oil revenue to build oil-based and other industries (including mechanised agriculture) in their recent plans.

The O.P.E.C. countries are also aware of the need to protect the real value of their earnings; to protect themselves against inflationary trends in the Western oil-consuming countries (who sell them machinery, etc., whose prices are rising with inflation), as well as against depreciations in the values of foreign currencies they hold. Their insistence on linking oil prices with an index of import prices is a reflection of their concern for guarding their earnings against such erosion of real prices.

The biggest asset of O.P.E.C. is the unity of its members. However, there are signs of serious divergences among them on fundamental policy issues. On the one hand, countries with diversified economies such as Iran intend to follow a policy of high prices which would allow for large revenues and foreign exchange earnings which they are capable of investing in the non-oil sector with the long-run objective of a sustained high rate of growth with reduced dependence on oil. On the other hand, desert economies (such as Saudi Arabia, Kuwait, or Abu Dhabi) with few resources other than oil and limited opportunities for diversification, are keen to maintain the competitive position of oil *vis-à-vis* other energy sources, in order to protect their future earnings from oil. For them, a high oil price generates reserves which cannot be put to immediate use, while it encourages the oil companies and oil consuming countries to research on cheaper alternative energy sources which, if successful, would jeopardise their future earnings. Given these basic differences, it is no easy task to evolve a common price formula for all types of oil producing countries. Policy differences also arise in view of wide variations in cost and production — reserve ratios among the oil producing countries. Some countries, such as Saudi Arabia, possess major oil reserves, which, at the current rate of exploitation, will last many years. Others, such as Venezuela or Algeria, are not so lucky with oil reserves, and would favour a conservationist policy.

The O.P.E.C. countries collectively gain from their unity. The difference unity makes can be seen from the achievements of O.P.E.C. in only one-and-a-half decades. On the other hand, like non-union workers in

Britain who benefit from trade union actions for wage increases while refusing to shoulder the responsibility of membership, some oil producing countries might find it advantageous to leave O.P.E.C. and offer oil at cheaper prices.

O.P.E.C. derives much of its strength from the political unity of Arabs during the aftermath of the 1973 war. But if the past experience is any guide, such a unity is fragile. There are serious ideological differences between 'left-wing' (e.g. Algeria, Iraq) and 'right-wing' governments (especially the Sheikdoms), and between Iran and Arab oil producers. Some of the smaller oil producing countries — such as Kuwait (pop. 800,000), Bahrein (200,000) and Abu Dhabi (86,000), with more than half of their population consisting of foreigners — are extremely suspicious of their neighbours who maintain territorial claims on these enclaves of prosperity. The multinational oil companies have always been adept at using these differences to play one country off against another, and thereby making their task of presenting a common front against the former a formidable one.

Furthermore, most of these regimes — feudal and autocratic as they are — are under the constant fear of revolution. Both for fighting the revolution, and for securing refuge if the revolution succeeds, these regimes are critically dependent on the support of the Western countries. This is despite the official 'anti-U.S.' rhetoric of the spokesmen of these regimes. While the forces of Arab nationalism and economics are pushing them towards strengthening the unity of O.P.E.C. countries, their resolve to remain united might weaken under the pressure of political and military considerations.

6.   The Poor Oil Consuming Countries (the 'nopecs')

What has been the impact of the high oil prices on the less developed countries of the world? It is easy to identify three direct, negative effects — in terms of balance of payments, foreign aid availability, and the problems created for their agriculture.

First, balance of payments. According to a rough estimate prepared by the World Bank, out of $65 million additional earnings of the oil producing countries in 1974 about $10 million would come from the poor oil consuming countries. This is a staggering sum which has put further pressure on the already precarious balance of payment situation of many of these countries. The hardest hit are the 630 million people of India, Bangladesh and Ceylon; and coming next in order are 90 million people of ten African countries: Ethiopia, Tanzania, Kenya, Malagasy, Cameroon, Upper Volta, Mali, Malawi, Rwanda and Somalia.

Not all the poor countries have suffered equally from oil price increases. For some, particularly those exporting mineral products, such as Zaire, there have been compensatory increases in the prices of their exports in recent years. But others have been affected by increases in prices of both oil and other new materials. Most of the affected countries have been forced to reduce their oil imports and other commodities; while frantically looking for 'soft loans' and sellers of oil who would accept deferred payment. Credit facilities offered by various international institutions have partly remedied this situation, but for most, the higher oil prices have meant a drop in their growth rates.

Secondly, the balance of payments situation has worsened for another reason; the amounts of grants and loans available from the Western countries have been curtailed, since the latter themselves are facing balance of payments deficits *vis-à-vis* the oil producing countries.

Thirdly, with the high oil prices, the new strategy for agricultural development with a new technology associated with high yielding varieties of seeds has now suffered a big setback. The new varieties have been designed to be responsive to chemical fertilizer; and being new, and often more vulnerable to attack from pests, they require the protection of pesticides; and since the use of fertilizer also encourages the growth of weeds, weedicide is another important component of this new strategy. The new seeds also require a controlled water supply (hence the need for pumps) and the use of tractors helps in ploughing the fields in time for the sowing of a second crop.

All these inputs are either oil-based (fertilizer, insecticide, weedicide) or are usually run with oil as motive power (diesel pumps and tractors). The increase of oil prices has led to a doubling of the prices of most of these inputs, with a consequent reduction in their use. Thus, for a large number of these poor countries the higher oil prices would seem to have made the struggle for matching population growth with increased food supply that much harder.

However, the negative arguments given above are in reality not as convincing as they sound at first. To take the balance of payment argument first, it will be wrong to put the entire responsibility for the balance of payments difficulties of these poor countries on the oil prices alone. The present balance of payments situation is the end product of a whole set of factors, including the doubling of prices of foodgrain exports, a major item of import for many poor countries, which mostly comes from North America. It is ironic that while the government of the United States expressed so much concern about high oil prices and their effect on the poor countries, and even went to the extent of organising a conference on oil importing countries in February, 1974; no such initiative was shown by them when food prices rose.

Again, while the drying up of foreign aid from the Western sources following the 1973 oil crisis is a fact; some would argue that it was a boon in disguise, since, given all the conditions and restrictions under which such aid becomes available, it favours the donor country perhaps more than the recipient. Furthermore, to the expert, the high oil prices have involved a large scale transfer of resources from the Western countries to the O.P.E.C., and given the latter's low marginal propensity to consume such resources in the short run, there is a clear possibility of a larger sum of money being available for distribution as aid. If the O.P.E.C. countries do assume the role of aid-givers seriously, this would reduce the dependence of the poor oil consuming countries in the West.

On the issue of 'green revolution', based on the new agricultural technology; while the oil crisis has no doubt affected its performance, some would argue that this strategy was bound to fail because of its 'rich-farmer' bias, irrespective of the impact of oil prices. In the case of India the 'green revolution' experiment stopped making progress after 1971, about two years before the big rise in oil prices.

Furthermore, while most of the poor countries have no doubt been affected in one way or another by the oil crisis, it is not always their poor citizens who have been the hardest hit. In fact the opposite is more probably the case, given the observed positive relationship between one's economic status and the consumption of energy. It is not the poor who drive automobiles, use synthetics, and own tractors. In India, half the total energy consumption is accounted for by non-commercial energy sources (e.g., cow dung, dry leaves etc.); and the most important oil product for domestic consumption is kerosene, an illuminant, a substitute for electricity. Unlike food prices, which affect the poorer sections more than the rich (given the Engel's law of declining percentage of income spent on food with increase in income), the incidence of oil price increases appears to be progressive.

On the positive side, if the precedent established by the O.P.E.C. countries — of forcing a shift in the terms of trade in their favour through collective action — is followed by other groups of primary producing countries, the resulting improvements in their export prices would more than offset the losses through high oil prices. Until the O.P.E.C. action, although there had been so much discussion in the literature on development about the terms of trade being favourable to the countries exporting industrial goods and importing raw materials, no concrete measures were taken, either by the international agencies or by the developed countries to move the terms of trade in the other direction. A lasting consequence of the O.P.E.C. move might be a restructuring of the pattern of world trade in favour of the poor primary producing countries.

A number of inter-governmental organisations of mineral producing countries already exist; for example C.I.P.E.C. (copper), I.B.A. (bauxite) and I.T.C. (tin), and recently exporters of iron ore met in Delhi to form their own organisation. But how far these organisations will succeed in following O.P.E.C.'s lead in increasing their export prices remains to be seen. Unlike oil which is exhaustible, most of these minerals can be recycled. Furthermore, most of these have several close substitutes in their uses; so in their case a policy of demanding higher export prices would work only if the exporters of a group of substitutable minerals band together.

Furthermore, many of the ill effects of the high oil prices can be removed by the oil producing countries themselves. In some cases the O.P.E.C. countries have been generous in providing credit facilities (e.g. India's recent deals with Iran and Iraq). O.P.E.C. countries have also taken initiatives in setting up development funds for the poorer countries. Occasionally there have been talks of a discriminatory price structure which favours the poor countries, although no action has yet been taken in this direction.

An important feature of the new phase in the world oil market is that it has created opportunities for direct dealings between these two groups of countries — O.P.E.C. and the poor countries — independently of the major international firms. The cooperation between them can take, and is in practice taking, many different forms, such as (a) jointly owning refineries with the crude oil being supplied by the O.P.E.C. partner; (b) jointly undertaking exploratory work in O.P.E.C. countries; and (c) joint operation of oil-based industries such as fertilizers and petro-chemicals. The cooperation can easily extend beyond oil and oil-based industries, and include transfer of technology, both skilled and non-skilled manpower, and so on. Above all, as we have already noted, if the O.P.E.C. countries agree to transfer a good part of their surplus oil earnings to the poor countries by way of investments and loans, not only would this compensate for the curtailment of loans and grants by the Western developed countries, such a step would lessen the economic and political dependence of the poor countries on the latter. From the point of view of the O.P.E.C. countries, this would be politically effective, both in creating goodwill in poor oil consuming countries and in diversifying their investment of oil revenue.

However, the sum total of the involvement of the O.P.E.C. countries in the affairs of the poor countries have so far been of very small magnitude compared with their resources. True that non-oil Arab countries have benefited from their help, as also politically important countries like India have; but by and large the less developed countries have been left in the cold by the O.P.E.C. While it is not difficult to find a line or two about the need to help the poor countries in the public statements of the representatives of O.P.E.C. countries, their main effort has been expended towards strengthening economic ties with the developed countries.

This is unfortunate, but not surprising. First of all, the risks of nationalisation and political upheavals are relatively higher in the poor countries, while the returns to investment are lower. Secondly the majority of these countries are politically and militarily integral parts of the Western politico-economic system; and hence it is logical for them to adopt measures which would strengthen rather than weaken the latter. And thirdly, as it is clear from their public statements, the main ambition of the Shah of Iran and his colleagues in the O.P.E.C. is to lead their respective countries into the 'rich man's club', rather than the salvation of the poor majority of the world. Whereas until recently the O.P.E.C. countries were treated as a part of the Third World of poor countries, the former now form a category of their own ('The Fourth World').

Footnotes

[1] These companies are: Standard Oil of New Jersey (Exxon), Socony Mobil Oil Company, Standard Oil of California, Gulf Oil, Texas Oil (all U.S. based); British Petroleum (U.K.); and Royal Dutch & Shell. For a detailed account see Biplab Dasgupta[1] 'Large International Firms in the Oil Industry', in *I.D.S. Bulletin*, October, 1974 (Special Issue on Oil and Development); see also Edith Penrose[1] *The Large International Firm in Developing Countries: the international petroleum industry*, London, 1967.

[2] By 1965 the percentages became 76%, 58% and 66%. See Zuhayr Mikdashi, *The Community of Oil-Exporting Countries: a study in governmental cooperation*, London, 1972.

[3] In this paper most of the issues have been considered from the point of view of the O.P.E.C. countries of the Middle East. Countries like Venezuela and Indonesia have not been given much attention because of their declining role in the world oil market.

[4] See Dasgupta[1] (*op. cit.*) and Penrose[1] (*op. cit.*) for details on points covered in this section.

[5] Biplab Dasgupta[2], *The Oil Industry in India — some economic aspects*, London, 1971, chapters 2—3.

[6] See Michael Tanzer, *The Political Economy of International Oil and the Underdeveloped Countries*, London, 1969.

[7] For details see Biplab Dasgupta[3], 'Soviet Oil and the Third World', in *World Development*, May, 1974.

[8] British Petroleum, *Statistical Review of the World Oil Industry*, 1972.

[9] See Petroleum Economist, May, 1974.

[10] Petroleum International, November, 1974.

[11] See H J Frank, *Crude Oil Prices in the Middle East: a study in oligopolistic price behaviour*, 1966.

[12] Dasgupta[2], *op. cit.*

[13] See W L Newton, 'The Longrun Development of the Tanker Freight Market', *The Journal of the Institute of Petroleum*, September, 1964.

[14] Edith Penrose[2], 'Profit-Sharing between Producing Countries and Oil Companies in the Middle East', *Economic Journal*, June, 1959.

[15] See Dasgupta[3], *op. cit.*

[16] Mikdashi, *op. cit.*

[17] By the late sixties all other major oil exporting countries (excluding U.S.S.R.) joined O.P.E.C.

[18] For details on the points covered in the next two paragraphs, see Mikdashi, *op. cit.* (chapter 2).

[19] For several years, from 1964 to 1971, the O.P.E.C. countries allowed a certain percentage of discount on posted prices for tax purposes. See Mikdashi, *op.cit.*, p. 143.

[20] Frank Ellis, 'Statistical Background', *I.D.S. Bulletin,* Oct. 1974, (Special Issue on Oil and Development), p.29.

[21] See British Petroleum, *op. cit.*

[22] M Adelman, *World Petroleum Market,* 1972.

[23] See Mikdashi, *op. cit.,* chapter 5, for reasons for failure.

[24] Petroleum International, August, 1974.

[25] Estimates prepared by the World Bank. See Richard Jolly, 'Assessing Economic Impact on Developing Countries and some Policy Suggestions', *I.D.S. Bulletin,* October, 1974.

[26] There was a rush for bilateral deals with Arab countries in the aftermath of the 1973 war. In January 1974 France, West Germany and the U.K. entered into agreements with Saudi Arabia, Algeria and Iran. See Frank Ellis, 'Diary of Events in the Oil Market 1971—74', *I.D.S. Bulletin,* October, 1974.

THE ROLE OF LESS DEVELOPED COUNTRIES IN WORLD RESOURCE USE

Paul Rogers

A Colonial Legacy

A common feature of past empires has been the transfer of resources from the periphery to the centre, the conquered peoples providing the raw materials and other resources needed to maintain the imperial power in its accustomed state. In the past, resources as diverse as grains, spices, salt, precious metals, gems and fruit have flowed to the centres of empires with, until very recently, slave labour being the most important resource of all.

Only in the sixteenth and seventeenth centuries did this phenomenon develop on a world wide scale with emergence of imperial powers with intercontinental influence. Among the first were the Spanish and Portuguese followed by the British, French, Belgians, Dutch, Italians and Germans. More recently still, other major powers have emerged such as the United States and Japan which, while lacking the more obvious trappings of imperial power, have controlled the exploitation of resources in many countries across the world.

The great imperial continent was Europe. Initially, conquests of semi-tropical and tropical lands were made as part of a quest for very high value resources such as gems, precious metals and spices. But by the early nineteenth century, several factors had combined to make the control and continued supply of much cheaper raw materials a desirable aim of the imperial powers.

The demands of the industrial revolution steadily consuming home-based supplies of raw materials, the growing populations of the newly industrialised countries and major improvements in the efficiency of freight transport by sea made it both necessary and possible to rely on distant colonies as providers of cheap raw materials.

The principal motive in the colonising process was the acquisition of wealth by the colonising powers. It is true that, in time, some of the excessive features of repression were ended, with reforming action such as the anti-slavery movement paralleling at an international level the zeal of reformers within the industrialised countries to overcome the injustices of phenomena such as child labour. But the essential feature of the colonising process was not the "betterment" or development of the colonised peoples, it was the provision of resources for use by the colonising powers. At different periods in history this was the prime feature of European colonisation of much of Africa, Asia and Latin America.

In the decade following the end of the Second World War, the process of dismantling the structures of the European empires was set in motion but while this process is nearing completion, the nature of the relationships between colonisers and colonised has, even now, changed little. At and after independence, many colonies retained very close ties, especially economic ties, with their former masters and inherited a situation where their main function in the world economic order was to continue as suppliers of raw materials, with the prices effectively dictated by the purchasers.

This was made more rigid by the fact that many countries lacked versatility in their range of exports to their former colonial masters, with export earnings coming from very few commodities. Thus Uganda produced cotton and coffee, Tanzania exported sisal and Zambia produced copper. For Ghana it was cocoa and for Sri Lanka, tea. Malaya provided tin and rubber and Pakistan produced jute. These examples from the British Empire could be repeated with examples from former colonies of many other European powers. Even the countries of Latin America which had won their independence many decades or even centuries earlier were still frequently restricted to one or two commodities for much of their export earnings. Tin from Bolivia, copper from Chile, coffee from Columbia and bananas from many Central American republics are examples of this situation.

In all, this colonial legacy provided little in the way of economic independance for the producing countries. A country dependant on one commodity for half its export earnings and on perhaps two commodities for as much as 80 per cent of such earnings was clearly very dependant on the vagaries of world market price trends for these commodities. A price slide could adversely affect development planning for years ahead.

Although these factors of commodity specialisation meant that a country might lack economic independence they did not mean that one or other commodity was necessarily produced by just a few countries. For some commodities this was true in that supplies of tin, copper, rubber and some other resources from third world sources involved very few countries. But other commodities such as tea, sugar, coffee, bananas and oil seeds were produced by many countries.

We thus have a world economic order in the early post-war years in which a great many less developed countries continued in their role as providers of cheap raw materials for the developed industrialised countries of the world. This was during the period of the consumer revolution when the differences in wealth between rich and poor countries widened. Indeed, one of the factors which ensured the continuance of the consumer revolution in developed countries was precisely this continued availability of cheap raw materials.

This circumstance can now be recognised as a pre-requisite for producer power. Why then did this power remain dormant until the early 1970s?

Some Aspects of World Trade

The nature of trade between rich and poor countries in the 1950s and 1960s remained fairly constant. Only the advent of producer power and the 1973/4 commodity price boom resulted in any change. Essentially it was a period of two decades during which the less developed countries maintained their position of suppliers of raw materials to the developed industrialised countries.

Even so, their share in world trade was small and declining and they had no more than a minority share in world trade in primary products. This may be surprising when one considers that we are concerned with countries making up the majority of the world's population, but it is some indication of the way in which relatively few wealthy countries dominate the world trading system.

In 1953, world exports totalled $78,000 million f.o.b., with the share of the less developed countries amounting to slightly over 25 per cent. By 1967, the value had risen to $214,000 million f.o.b. with the less developed countries' share having fallen to about 19 per cent. Table 1 below gives the exports of the less developed countries by destination for 1953, 1960 and 1967. This shows two interesting aspects of their trade. Firstly, it was primarily with the developed countries rather than with each other and secondly, while the amount of trade was increasing, the amount of trade among less developed countries declined relative to that with developed countries[1].

Concerning the categories of exports from developed and less developed countries, in 1967 primary products accounted for about one quarter of the value of exports from developed countries, amounting to approximately 35,000 million dollars in value. In the same year primary commodities accounted for nearly 80 per cent of exports from less developed countries, but the total value of their exports was so much lower than

Table 1 Exports of less developed countries by destination

	Per cent of less developed countries' exports		
Destination	1953	1960	1967
Developed countries	71	70	72
Centrally planned countries	2	4	6
Less developed countries	25	23	20
Total value (thousand million dollars)	21	27	40

that of the developed countries that the value of the primary commodity exports was only 32,000 million dollars, less than that of primary commodities exported by the developed countries.[1]

Regarding direction of trade, most of the primary commodities exported by less developed countries went to developed countries, whereas the former imported mainly manufactured goods from the latter. Thus the pattern of trade for the less developed countries was one of exporting primary commodities and importing manufactured goods, the great majority of both kinds of transaction being with developed countries.

Following a short period of relatively high commodity prices in the early 1950s, many less developed countries experienced a decline in their terms of trade (ratio of export prices to import prices) over a period of nearly two decades. Generally speaking, commodity prices during this period rose more slowly than the prices of manufactured goods. This applied to petroleum exporting countries and to other developing countries although the former redressed the balance in 1971 and 1972 by effecting modest increases in export prices. Table 2 shows, however, that this did not extend to other developing countries.[2] Indeed according to one UNCTAD estimate, by 1972

"the terms of trade of these countries had deteriorated by about 15 per cent, compared with the mid-1950s, equivalent to a loss, in 1972, of about $10,000 million, or rather more than 20 per cent of these countries' aggregate exports, and considerably exceeding the total of official development assistance from developed market economy countries to developing countries in Africa, Asia and Latin America (some $8,400 million in 1972). In other words there was, in effect, a net transfer of real resources, over this period, from developing to developed countries, the flow of aid being more than offset by the adverse trend in the terms of trade of the developing countries."

Table 2   Per cent changes in export and import prices and in the terms of trade
of developing countries, 1954 to 1972

Period	Petroleum-exporting countries			Other developing countries		
	Export prices	Import prices	Terms of trade	Export prices	Import prices	Trades of trade
1954-6 to 1968-70	+ 2	+ 13	− 10	− 1	+ 12	− 12
1968-70 to 1972	+ 39	+ 17	+ 19	+ 12	+ 17	− 5
1954- 6 to 1972	+ 42	+ 33	+ 7	+ 11	+ 32	− 16

In summary then, the less developed countries of the world had a
relatively minor role in world trade in the period up to 1972. Their role
was one of providers of raw materials to the developed countries,
especially the developed industrialised market economy countries.
During an 18 year period up to 1972, their position in world trade
declined in two senses, by virtue of their declining share and by virtue
of their declining terms of trade.

The Commodity Boom

This prevailing feature of world trade altered slightly for a brief period
in the early 1970s when a price boom affected a large number of commodities originating partially from third world sources. The boom
developed during 1973 as can be shown by Table 3 which covers world
commodity price increases excluding oil.[3]

Table 3 World Commodity Price Increases during 1973

Item	Price at 12 December 1973 (1963 = 100)	Per cent increase in 12 months
Food	262.5	36.8
Fibres	239.7	59.5
Metals	437.8	133.0
Miscellaneous	265.7	70.3
All items	272.6	60.2

There were a number of causes of this commodity price boom, the
most important being a period of synchronised industrial expansion in

a number of developed industrialised countries, but lesser factors included a better showing for some commodities normally likely to experience competition from synthetics with a petrochemical base. These were subject to fears of price increases by consumers following the OPEC-instigated price rises for oil in September 1973. The commodity price boom reached its peak for most commodities during 1974 and there were then sharp declines in prices by early 1975, as shown in table 4.[4]

Table 4 Commodity Price Trends, 1974—1975

Commodity	Currency and unit	1974 peak price		April 1975 price	Percent price decrease from peak
Sugar	US cents/lb	56.63	(Nov)	24.06	58
Coffee	US cents/lb	73.74	(May)	59.53	19
Cocoa	US cents/lb	82.74	(May)	56.22	32
Tea	New pence/kg	66.76	(Mar)	60.20	10
Cotton	New pence/kg	97.81	(Jan)	48.57	50
Sisal	US $/metric ton	1070	(Jun)	756	29
Abaca	US $/long ton	907	(Jan)	523	42
Rubber	US cents/lb	49.2	(Jan)	24.4	50
Copper	£/metric ton	1268	(Apr)	561	56
Lead	£/metric ton	303	(Mar)	202	33
Zinc	£/metric ton	738	(May)	303	55
Tin	£/metric ton	3951	(Sept)	3007	24

Even though the commodity boom was considerable in extent, it did not greatly alter the position of the less developed countries in the world trading system. Hone, for example, estimated that their share of world trade improved from 17% in 1972 to no more than 19% in 1973.[3] In addition to this, the boom coincided with the oil price rises and with accelerating inflation of the prices of manufactured goods imported by less developed countries from developed industrialised countries. In addition to this the benefits were highly variable on a country to country basis. Some major producers of minerals did well for short periods as did sugar producers and growers of some fibre crops and rubber. But many countries exporting those commodities which were relatively unaffected by the boom were hit hard by the increased costs of imports. Especially affected were countries with large populations such as India, Bangladesh and Pakistan.

For even those commodities showing major price increases, the real situation was often less than satisfactory. Producers of natural rubber, for example, did benefit indirectly from oil price increases but the 1974 average price for rubber was still 24% lower than the average price for the 20 year period 1953—72 (in constant 1974 US dollars). In April 1974, a little after the January high, its price was down 46% on the

1953—72 average.[4] This highlights the importance of taking into account the effects of inflation. Commodity price variations over a number of years, provided they are given in terms of constant currency valuations, all too often show a disturbing picture of stagnation, even during an apparent boom period.

Another example is jute which reached its peak only at the beginning of 1975. In 1974, for example, the average price of jute was still 30 per cent below its average price for the 1953—72 period (again in constant 1974 US dollars). Finally, tea, a very important commodity for countries such as Kenya and Sri Lanka hardly shared at all in the commodity boom. Its price, in real terms, has been following a downward trend during the past two decades, the 1974 average being 41 per cent below its 1953—72 price.[4]

In addition to all this, of course, the increased prices of many raw materials have had a direct effect on less developed countries because they, just like the developed industrialised countries, import raw materials, albeit on a much smaller scale. In this context, though, the greatly increased costs of food grains have had a marked effect in many less developed countries.

In summary then, the role of the less developed countries in trade in primary commodities has been one of supplying these commodities to the developed industrialised countries of the world. In the 1950s and 1960s their position in world trade deteriorated and their terms of trade declined. A short-lived commodity boom in 1973 and 1974 partially reversed these trends for some but not all less developed countries, but only on a temporary basis, and the first few months of 1975 saw a marked slump in the prices of most primary commodities.

Producer Power

Against this background of poor trade prospects it becomes possible to appreciate that there are many obstacles in the way of the exercising of producer power by less developed countries. Amongst the more important requirements for producer power are

(a) that the third world countries producing a particular commodity have a high degree of unity of action,

(b) that they control a large proportion of world exports of the commodity in question,

(c) that, if the commodity is based on a non-renewable resource, they control a large proportion of the world's known reserves,

(d) that large stockpiles are not held by consuming countries,

(e) that substitution by other commodities or by synthetics is not feasible, at least in the short term, and

(f)   that the countries seeking to exercise producer power should be sufficiently strong to resist any pressures which may be put on them by consumers.

It is interesting to note that, in the case of oil, (discussed in detail by Dr Dasgupta) all these requirements were fulfilled by OPEC. OPEC itself, although composed of nations with greatly differing political outlooks, had acquired a considerable degree of unity in its 13 years of existence prior to October 1973. The OPEC member states were responsible for the majority of world oil exports and held approximately 60% of known reserves of crude oil. Most consuming countries held stockpiles providing less than 100 days supply and alternative sources of energy were not available in sufficient quantities in the short term. Finally, the OPEC member states had already amassed considerable foreign exchange reserves, a valuable deterrent to any consuming country seeking to take drastic action to ensure security for its oil supplies.

We may now turn to examine some other commodities and their potential for producer power. The first point to make is that no other important commodity would appear to have the potential of oil for the exercising of producer power. If we look at minerals first, table 5 gives some relevant data for those minerals which might be amenable to commodity bargaining by third world countries.[5]

Table 5

Resource	Less developed countries with major reserves (% of world total)		Major producers among less developed countries (% of world total)		Estimated life of global reserves in years
Aluminium	Guinea	(20)	Jamaica	(19)	31
(as bauxite)	Jamaica	(10)	Surinam	(12)	
Cobalt	Zaire	(31)	Zaire	(51)	140
	Zambia	(16)			
Copper	Chile	(19)	Zambia	(13)	21
			Chile	(10)	
Nickel	Cuba	(25)	New Cale-		
	New Cale-		donia	(28)	140
	donia	(22)			
Tin	Thailand	(33)	Malaysia	(41)	15
	Malaysia	(14)	Bolivia	(16)	
			Thailand	(13)	
Tungsten	China	(73)	China	(25)	26

Several aspects of the table have to be treated with reservation. The figures for reserves are tentative and are continually changing and the figures for estimated life of global reserves are, of course, approximations based on extrapolations of existing trends. Their value is to serve as a reminder that mineral reserves are non-renewable and *do* get depleted. New reserves may be discovered but they may tend to be more difficult to exploit. In any case, the exercising of producer power does not have to await reserve depletion. This was certainly not the case with the OPEC price rises which commenced at least two decades before the likely depletion of Middle Eastern oil reserves.

On a slightly different note, the column in table 5 concerning mineral production must be treated with caution. A less developed country may produce only 10 per cent of the world supply of a mineral but virtually all of that will be exported. Many developed countries may have their own supplies yet still not be self sufficient. A less developed country with a minority of world production may therefore dominate world *trade* in that mineral, trade being the important factor for producer power.

Even so, there are few minerals in which less developed countries have a dominant role in world markets to the extent of the role of OPEC for oil. Furthermore, there are considerable possibilities for substitution (eg. aluminium and copper) and replacement with synthetics. Stockpiling is possible for many minerals, particularly those with a high value per unit weight, and, finally, the financial strength of any third world mineral producer rarely approaches that of even one of the weaker members of OPEC.

Regarding agricultural resources, the theoretical prospects for producer power look at least as unpromising. With a few exceptions such as rubber and sisal there are usually a large number of third world countries producing a particular agricultural commodity. Coffee, cotton, tea and sugar are examples of this. Synthetic substitution is possible for most fibre crops and may become increasingly likely for beverages and food crops.

On the other hand, many agricultural commodities have a greater degree of supply rigidity than is often conceded. Many plantation crops may take 1 — 5 years to come into bearing from the time of planting, making it less easy for consumers to utilise alternative supplies. The greatly increased price of oil has made synthetic substitutes with a petro-chemical base less competitive with natural products and, finally, there are many agricultural commodities from tropical countries which cannot readily be replaced by synthetics and which cannot be produced under the temperate conditions existing in most developed industrialised countries. But even these factors cannot easily outweigh the problems facing producers of agricultural commodities.

Taking into account these various considerations, it must be admitted that the prospects for less developed countries to exercise producer power are rather limited. Yet it is plainly obvious that in the last three years there has been a surge in activity directed precisely towards this end. The reasons for this apparent contradiction will be examined in a later paper and tend to lie outside the field of world trading matters, extending into the spheres of politics and international relations. The aim here is to review, briefly, some of the examples of these moves towards producer power by third world countries exporting commodities other than oil.

Phosphates

After 20 years of stagnation of rock phosphate prices, world overproduction came to an end in 1973 and in that year demand exceeded supply by about 5 per cent. Morocco is estimated to control about 50 per cent of total world reserves of rock phosphate and its annual exports of 16 million tons represent 16 per cent of total world production and a very much larger proportion of world trade in rock phosphate. On 1 January 1974, Morocco doubled the price of its rock phosphate exports and on 1 July 1974 it increased the price of a further 50 per cent representing a tripling of the price in six months.

Morocco felt it had some justification in effecting this price increase. In 1973, the world selling price of one of the best grade of phosphate, 75–77 per cent Bone Phosphate of Lime from the Khourigba workings was 14.15 dollars per metric ton. Twenty-one years earlier in 1952, the price had been 14.20 dollars.[6] Taking into account inflation over that period, the Moroccan view was that the price increases were redressing what had been a worsening trading position.

The case of phosphates was the one other example of successful producer power in the 1973–1974 period apart from the oil price rises. It was possible because of a shortage of phosphates coupled with the fact that one country dominated the world export market.

Copper and Aluminium

These will be considered in depth in a later paper in this volume, and mention will only be made here of some aspects of the two producer groupings now in existence.

About 37 per cent of world copper reserves lie in the developed countries of the western bloc, a further 16 per cent are controlled by centrally planned countries while 47 per cent of reserves are found in less developed countries.[7] Four such countries, Zaire, Zambia, Peru and Chile form CIPEC, the Intergovernmental Committee of Copper Exporting Countries. CIPEC, which has been in existence for a number of years, cannot parallel OPEC's record of success, not the least of the problems being political disunity and the more practical matter of copper's possible substitution in many of its roles by aluminium.

However, in the face of falling copper prices in the latter part of 1974, CIPEC member states agreed to cut output by 10 per cent in November 1974 and by a further 5 per cent early in 1975. In July 1975 CIPEC announced that its members had carried out this proposal, exports for the five months ending in April 1975 were 885,000 metric tons against original plans for 987,000 tons.[8] Earlier in the year there was a report that Australia was expressing interest in joining CIPEC and that overtures were being made to two other countries, Papua New Guinea and the Philippines.[9] More significant was a report a few weeks later that Iran was being approached as a possible member[10] in view of the exploitation of its southern deposits, but also a development which might improve links between CIPEC and OPEC.

The International Bauxite Association was set up early in 1974 and it is too early to say whether it will achieve any degree of effective producer power. IBA includes developed as well as less developed countries as Australia and Yugoslavia are members. The production and processing of bauxite is a complex matter with costs varying greatly from one centre to another. This makes common pricing policies difficult to manage, but new processes for the extraction of aluminium from sources other than bauxite are likely to become available within a few years and this may spur IBA to speedy action.

Iron Ore

The great majority of the world's iron ore production comes from developed industrialised countries with the Soviet Union (28 per cent of world production) and the United States (12 per cent) leading the field.[11] Nevertheless, very few industrialised countries are self-sufficient and many import at least a minority proportion of their requirements from less developed countries.

The initiative to form an iron ore producers' association came from India late in 1974 and by mid-January 1975, 11 countries had met in New Delhi and had agreed to form an Association of Iron Ore Exporting Countries. The group was made up of Algeria, Australia, Brazil, Canada, Chile, India, Mauritania, Peru, Philippines, Sweden and Venezuela, thus including three developed countries.

It was agreed that the association would come into being after seven participants had signed the agreement establishing the association. Brazil and Canada refused, at first, to sign as their delegates expressed the view that consuming countries should be involved in the association. However, by 15 September 1975 seven countries had signed the agreement (Algeria, Venezuela, Australia, India, Chile, Peru and Mauritania) and AIOEC came into being with headquarters in London.[12]

Natural Rubber

An Association of Natural Rubber Producing Countries (ANRPC) includes Malaysia, Thailand, Indonesia, Sri Lanka, South Vietnam and Singapore, with the first three controlling some 85 per cent of total world output of natural rubber. Natural rubber makes up only 30 per cent of world rubber consumption, synthetic forms predominating not least because of severe price fluctuations for natural rubber.

In April 1974, Malaysia, Indonesia and Thailand agreed to work together to produce a plan to stabilise prices through buffer stocks and possibly direct production controls. Price stabilisation is probably a necessary preliminary to any more radical action aimed at increasing the price of natural rubber. This itself may become more feasible with increased prices for synthetic rubber due to oil price rises.

By August 1975, the group within ANRPC working on these proposals had expanded to include Singapore and Sri Lanka and had reached a large measure of agreement on the nature and financing of the planned stockpile.[13]

Other Commodities

There have been moves towards increased cooperation among producers of many other commodities in recent months. Early in 1975, coffee producers established a six nation working group to study a proposal by Brazil and Columbia on the structure of a new international coffee agreement,[14] a significant step in the trading of a commodity in which Latin American and African producers have shown little sign of agreement in the past.

In April 1975, 22 sugar producing nations from Latin America and the Caribbean reached agreement on presenting a united front at the meeting of the International Sugar Organisation the following month. The significance of this move was that the nations concerned accounted for more than half of world exports of sugar.[9]

An unusual grouping of rich and poor countries met in April 1975 to set up an organisation of mercury exporting countries. The association, with headquarters in Geneva, would involve Spain, Italy, Yugoslavia, Algeria, Peru and Turkey. At an earlier meeting held in Algiers, potential member countries had agreed that a "reasonable" price for mercury would be approximately 5 dollars a pound compared with the free market price of about 2 dollars.

Finally, an association of tungsten exporters has been established[4] and an organisation of banana producers seems probable.

Conclusions

It is now evident that an increasing number of less developed countries see the exercising of producer power as a means of improving their export earnings. Until now, and with the exception of oil and rock phosphate, the moves towards producer power for other commodities have involved little more than the establishment of organisations. It is far too early to suggest the likely outcome of the activities of these new organisations, but one point is worthy of mention.

From the standpoint of the conventional view of the world trading system, the odds would appear to be against successful commodity bargaining by most third world producers of primary commodities. Contrary to this view though, producers show every sign of attempting to exercise producer power. The point to be made is that we are *not* here concerned solely with economic affairs. The success of the OPEC negotiations in 1973 and 1974 have shown a possible way for other third world resource producers. It is possible that producer power will evolve into a movement which is as much political as economic in nature and may be concerned with a re-aligning of international power blocs. Certainly the most important requirement for the successful use of producer power over the next few years will be international co-operation among the nations of the third world. The evidence of the last two years suggests that such cooperation, if not probable, is certainly possible.

Notes and References

[1] Table 1 and figures in text taken from "World Development Handbook" by Juliet Clifford and Gavin Osmond, Overseas Development Institute/Charles Knight Ltd., London, 1971.

[2] Table 2 adapted from "Problems of Raw Materials and Development" UNCTAD Document TD/B/488, UNCTAD, Geneva, 1974.

[3] "Gainers and Losers in the 1973 Commodity Boom: Developing Countries Prospects to 1980" by Angus Hone, ODI Review, No 1, 1974. London.

[4] Information for Table 4 and text discussion supplied by courtesy of the Commodities Division of UNCTAD.

[5] Data for Table 5 from three sources:
P R and A H Ehrlich, "Population, Resources, Environment". W H Freeman, 1972.
"Optimistic Report on World Resources" Financial Times, 21 March 1974.
D L and D H Meadows, "Towards Global Equilibrium: collected papers", Wright-Allen Press, 1973.

[6] "Survey of Morocco," Financial Times, May 1974.

[7] "OPEC as a Model for Other Mineral Exporters" by Cres Barker and Bill Page, in Institute of Development Studies Bulletin, Volume 6 Number 2, October 1974, Sussex, England.

[8] World Commodity Report, Number 28, 23 July 1975.

[9] World Commodity Report, Number 17, 29 April 1975.

[10] World Commodity Report, Number 25, 25 June 1975.

[11] From "The Politics of Scarcity" by Philip Connelly and Robert Perlman, Oxford University Press, 1975.

[12] Financial Times, 16 September 1975.

[13] World Commodity Report, Number 33, 27 August 1975.

[14] World Commodity Report, Number 15, 16 April 1975.

PROBLEMS OF MINERAL SUPPLY

F E Banks

It is almost an article of faith within a fairly large slice of informed opinion that we are entering into a period in which much of the world will be faced by a scarcity of non-renewable minerals. The oil crisis has been instrumental in making this a vital issue; although the Club of Rome report and its offshoots must be given their due. However, a global shortage, in the sense of having to face a situation where this or that natural resource is definitely extinct, is a long way off. In the short run, or so the theory goes, the prospect is for some kind of limited economic warfare, brought about by attempts by certain of the less developed countries to exploit their position as important producers of primary commodities. The machinery for doing this would be cartels on the pattern of OPEC; and perhaps even multi-cartels, in which producers of different raw materials come together for the purpose of forcing up the price of their products.

The stock of raw materials that can be found within just the upper crust of the earth is, in point of fact, tremendous. Moreover, up to now the international economy has generally functioned in such a way as to ensure that enough of these materials have been available to allow the industrial countries a trend growth rate of about two or three percent over much of the last century. The avowed purpose of cartels, or pro-ducers' associations, is to transfer some of the benefits that the industrial countries have been able to realize from processing these commodities to the producing countries. In order to justify the forming of these organisations, some rough concepts of social justice have been intro-duced into the discussion by various individuals and interest groups. These concepts, for the most part, turn upon the idea that the wealth of the industrial countries is possible only because of the extreme poverty of many of the producing countries.

In a sense this is true — though not in the sense that the reader probably believes. It is a simple, but not widely appreciated, fact that at the present level of mineral retrieval technology, the amount of either money or physical capital available for the world's mining sector is incapable of transforming even the most highly accessible portion of the known reserves of raw materials into flows of a magnitude sufficient to elevate the standard of living of more than a fraction of the world's population. Moreover, even if capital was available, the productivity of the underdeveloped countries outside the primary commodity sector is so low that the majority of the processing of these minerals would have to be done in the industrial countries — something which, again, would require additional amounts of capital, as well as astronomic amounts of energy, trained manpower, and so on. Under the circumstances it seems reasonable to postulate that the principle demanders of raw materials in the foreseeable future will continue to be those countries able to pay for the increasing difficulty of obtaining and processing them: in other words, the traditional industrial countries, and those few developing countries able to generate physical and social energies considerably greater than those in evidence in most of the third world today. The majority of the underdeveloped world will probably not press on the supply of raw materials for at least a century. Thus, when the demand for raw materials is spoken of in this paper, what is meant is a demand attributable to the customary sources, and evolving by a few percent per year. Quantum leaps, due to so-called Rostovian takeoffs into self-sustaining growth by large sections of the underdeveloped world, are considered to be beyond the realm of possibility at the present time.

Where the supply of raw materials attributable to the non-industrial countries is concerned, tables 1 and 2 should be observed.

Table 1. Supply of Selected Raw Materials from Non-industrial Countries

Metal	Percentage of:	
	World Output	World Reserve
Nickel	2	22
Lead	12	21
Zinc	13	23
Copper	34	47
Bauxite	43	57
Iron Ore	24	29
Tin	69	79

Table 2. Raw Material Production by Less Developed Countries[1]

	1950	1960	1970	1980[2]
Less Developed Countries	1.8	3.5	5.3	10.3
World	5.7	10.4	17.0	27.3

These figures indicate that the less developed countries provide the industrial world with about a third of their most important industrial raw materials. Some question might be raised as to just what change in this share could be made if such enterprises as President Gerald Ford's 'Project Independence' were to suddenly become the fashion in the industrial world. There are of course huge uncovered reserves of almost all the key minerals in Canada, Australia, Alaska, under the seas and oceans, and so forth; and in addition technology as it is developing should be able to extract minerals from ores that are considerably thinner than those being processed today. However at the same time it seems certain that even larger and more accessible reserves are available in various parts of the third world: South of the Amazon, in the Nubian-Arabian Shield, in various parts of Africa, etc. Moreover, given the fact that most of the financing of the industrialization, if any, of many of the less developed of the producing countries will have to be paid for by their receipts from the sale of raw materials (and in addition these receipts will have to cover a part of any increase in consumption they have planned), the chances are that we will be seeing an increase in the production and export of raw materials from these countries. Certainly it is unlikely that the near future will see any decrease in the absolute amount produced by the poorer of these countries. Instead, given the opportunity, they will attempt to expand their production as fast as possible.

If some of the assumptions advanced above do not correspond to those drafted by the committees of eminent people responsible for such chimeras as the United Nations' "First and Second Development Decade", or the "New World Economic Order", the reader should rest assured that the deviation is intentional. For instance, when I indicate that those underdeveloped countries furnishing raw materials to the industrial countries will probably continue in this role for an indefinite period in the future, I consider that I am merely expressing what any normally intelligent human being would regard as a certainty if he devoted just a few minutes thought to the subject. Put another way, I rule out a reversion of the traditional logic of economic development that would permit the Third World to continue its present rate of population growth, and its aggressive lack of attention to technical education and industrial discipline, and yet be able to initiate the type of industrialization that would allow it to become a serious competitor for the supply of raw materials.

Oil and Minerals

In examining the possibility of the industrial countries having to face more cartels of the OPEC type, the following point should be understood. If the producers of copper, iron ore, and the rest should meet one

fine evening and decide that the price they will require for these commodities shall be two or three times higher, then consumers of these
products will simply face a higher price the next morning. (This is
essentially what happened in the case of oil, and it has also happened
with bauxite from Jamaica, and phosphates from Morocco). Given this
situation, it seems reasonable that some attention be paid to what
countermeasures can or should be taken to bring these prices down.
For instance, at the beginning of the 'energy' crisis, some economists
argued that in due course, so called 'free market forces' would bring
about a reduction in the price of oil, and therefore it would be unneces-
sary for governments to resort to a deliberate application of various
policy devices. In a sense they were right: the *real* price (or purchasing
power) of oil has been decreased by almost one third, thanks to a combination of the demand reducing effects of large scale unemployment in
the industrial countries, and record price rises. If linearity is the rule in
these matters, then to return the price of oil to the level which some
economists have stated is suitable — which is one or two dollars a barrel
— would require about half the working force of the industrial world
to be collecting some form of welfare payment, as well as triple digit
inflation.

On the other hand, it would be a mistake to think that all the elements
of the oil situation are relevant to the analysis of non-fuel minerals,
since oil is unquestionably the most important industrial input of all
the minerals. There are several ways to see this, but the simplest is just
to look at the total value of some imports from the lesser developed
countries into the industrial world. In 1970, before the rise in oil prices,
oil was imported for a value of about 13 billion dollars. As for the total
value of production of the nine important non fuel minerals (copper,
bauxite, iron ore, manganese ore, nickel, lead, phosphorous, tin and
zinc), this amounted to only about 5 billion. Moreover, at the present
time, the value of oil imports is somewhere in the range 50–75 billion;
while the corresponding value of production of these commodities has
increased to about 7 billion. What this means is that the danger is not
from cartels or — as will be argued below — moderate price increases,
but arbitrary and for the most part unexpected price rises of the order
of two, three, or four hundred percent. The countermeasures mentioned
above should not be designed to provent the raw material producing
countries, and particularly the working members of the extractive industries in these countries, from reaping financial rewards as great as their
colleagues in the industrial countries, but to keep bureaucrats and diplo-
mats from being able to make careers for themselves rigging the prices of
vital raw materials, and rigging them in such a way as to promote
economic chaos or political extremism in the industrial world.

Before going to some individual commodities, it might be a good idea
to take a brief look at some aspects of the economics of exhaustible

resources.[3] We know that an unmined mineral is a stock which, if used today, is unavailable tomorrow. The maximization of profit (or surplus) thus requires that revenue obtained in the present period must not only cover marginal costs, but also the discounted value of unit profits given up by extracting minerals now instead of later. Put another way, a mineral should be left in the ground as long as its rate of profit appreciation is greater than the profit appreciation that would be realizable if it were extracted today and its profit allowed to grow at the bank rate of interest. Consider a simple numerical example where the profit from extracting a mineral in the present period is 100, and the *expected* profit from extracting it in the next period is 105. This is a profit appreciation of 5 percent. If the rate of interest was 10 percent, and the mineral was extracted now and its profit (= 100) put in a bank, it would grow to 110 in the next period. Thus it should be extracted at the present time.

This type of reasoning applies almost without modification to such important raw materials as copper, tin, lead, zinc, and other raw materials for which the general belief is that the amount of reserves available will be exhausted in the foreseeable future. We know that as reserves become scarce relative to demand, and the cost of extracting them increases, their price should show a tendency to rise. Thus if consumers want them at the present time, they should be required to pay a price large enough to compensate the producers for the profit that would have been realised had they left these resources in the ground. Moreover, having to pay the higher price at the present time hinders the over-investment in machines, structures, and durable goods whose profitability or utility would decline rapidly once the 'exhaustibility' of these resources was revealed by a rise in their price. An illustration applicable here is the over-investment in automobiles and buildings having a very high consumption of fuel, which was encouraged by the low level of fuel prices that prevailed until the October War in the Middle East.[4]

Given the fact that many minerals can be considered exhaustible at the present level of extractive technology, their optimal intertemporal allocation probably requires some sort of annual price increase. What these price increases would do is not only hold down present consumption, but also make profitable the development of substitutes, stimulate exploration, and so on. For example, had the price of oil increased slowly over the last decade or two, the technology for producing oil from coal, shale, tar sands, and the rest would be considerably more advanced by now. Interestingly enough, it may be the case that the present *real* price of about 8 dollars/barrel is only slightly higher than the optimal price. The problem is that it reached this level almost instantaneously, instead of gradually over a period of 10 or 20 years, and thus the economies of the industrial countries were unable to make the necessary adjustments.[5]

Copper

The next step is to examine, in some detail, the copper market. Prior to the early 1960's, the history of copper prices and the future of copper was mostly a matter of concern for the directors and stockholders of the large copper-producing companies. The buyers of copper had, with the exception of the World War II period, usually managed to get all the copper they wanted at reasonable prices; and outsiders such as economists were generally inclined to treat the copper industry as unworthy of their attention.

All this was changed by the nationalisation of many of the copper companies that began in the 1960s. Prior to these nationalisations the burden of a collapse in the price of this or that raw material fell on the firm producing it. With the nationalisations this burden was transferred to the countries in which this producing capacity was located. Thus although microeconomists continued to ignore the functioning of the copper industry, development economists began to give it some attention, and various UN organisations decided that it might be in their interest to "study" this commodity.

We can now look at the price of copper for the period 1950—73.

Table 3

Price of Copper (US Cents/Pound)

1950	24.0	1962	30.6
1951	24.2	1963	30.6
1952	24.2	1964	31.9
1953	28.8	1965	35.0
1954	29.7	1966	36.2
1955	37.5	1967	38.2
1956	41.8	1968	41.8
1957	29.6	1969	47.5
1958	25.8	1970	57.7
1959	31.1	1971	51.4
1960	32.0	1972	50.6
1961	29.9	1973	62.0

The upward movement of copper prices, which a number of people still prefer to think of as definitive, began about 1968 with a major strike in the United States. The prices shown above are average prices, but what happened was that in 1968—71 the price exceed 50 and even 60 cents/pound for fairly long periods, and thus gave the impression in certain circles that copper was on its way to becoming a rare metal.

As the war in Vietnam began to wind down, however, some of the steam seemed to leave this market. But at almost the same time the international business cycle turned up; and turned up in such a way as to provide a strong new impetus to the demand for copper. Even so, underlying expectations during this period pointed to a return to normality on the world commodity markets — that is, a situation in which the basic metals and other primary commodities would be available for prices that were, in comparison with other factors of production, relatively low.

There was also a problem with speculation — not so much with copper, but with the currencies for which much of copper was traded: dollars and pounds. In particular there had been an enormous increase in the international circulation of dollars due to the budget deficits that were used to finance the war in Vietnam; and the value of American currency was beginning to show an irreversible tendency to deteriorate vis-a-vis other currencies. This caused, among other things, an extensive speculation against the dollar in favour of some other currencies, and to a certain extent primary commodities. (Particularly through the 'futures' markets.) This speculation did not reverse the opinion of many analysts as to what was going to happen to the price of copper in the long run. But it did set the market on edge, and in addition reinforced some incorrect impressions of how the world economy was going to function in the future that were held by various economists who were becoming interested in the commodity markets, and who were in a position to relay these impressions to the general public via the popular press and television.

It was, however, the aftermath of the October War in the Middle East that introduced a kind of turmoil into the copper market. There are a number of reasons for this, most of them having to do with the psychological state in which buyers found themselves after having experienced the oil boycott. There was also the matter of copper producers in the less developed countries having formed a producers' association called CIPEC that, at least to many outsiders, took on some of the appearance of OPEC. What happened then was that copper purchasers in several countries, in particular Japan, were stampeded into increasing the size of their inventories of copper far beyond their requirements for current production. (These inventories, incidentally, still "overhang" the market, and if they were unloaded in a fairly short period of time, could be expected to reduce the price of copper by a fairly large amount.) In any event, these purchases drove the price of copper to record heights; and it was only the arrival of the demand-depressing effects of oil price rises and major unemployment in the industrial countries that checked this escalade in price.

The attitude of the CIPEC countries during this period was not without interest. As the price of copper increased, the governments of these countries, their 'experts', publicists, and well-wishers were able to attach a value to this mineral's importance that was completely out of line to the realities of the situation. It was at this time that the talk of CIPEC emulating OPEC reached a kind of fever pitch — although, to be fair, much of this talk came from know-nothing journalists in the industrial countries. Later, when the prices of raw materials began to collapse, with copper leading the way, CIPEC members began to think of price control as something to prevent the bottom from falling out of the market, and their economies from being ruined.

The question of just what the price of copper should be is difficult to ascertain. In my own work, beginning several years ago, I came to the conclusion that the normal price of copper, at the present time, should be about 450 pounds/ton, which is about 50 cents/pound. This figure, which I first published when the price of copper was about 800 pounds/ ton, and last published when the price of copper was 1200 pounds/ton, is intended to be reckoned in constant (deflated) pounds, and probably should be adjusted slightly to take into consideration some of the 'exhaustibility' effects mentioned above, and to a certain extent inflation.[6] It should be carefully observed, however, that this 450 pounds/ ton is estimated on the basis of the cost curves of the major producers, and involves existing capacity. (When the present downturn has run its course, and new capacity becomes necessary, it may be the case that this new capacity will not be forthcoming if the price of copper is not greater than the above estimate. Or so the story goes.) At the present time (May 1975) the *market* price of copper is 530 pounds/ton; and if this is deflated by an index that takes into consideration the current purchasing power of copper, it places the real price of copper at considerably less than 450 pounds/ton; but even so investment in new plant and equipment appears to be continuing at almost a normal pace.

We are now in a position to open our inquiry on the capabilities of the producers in the less developed countries to effect major increases in the price of copper. To begin with, we can consider the question of just what price these producers desire. In the discussion of equity between producer and consumer that constitures a large part of the so called 'work program' of the United Nations, and to a certain extent the CIPEC secretariat, it must be remembered that the rate of profit never comes into consideration, only the movement of prices. However given the fact that the wages of workers in South America and Africa are at most one-third of those in North America and Canada, and that seams are much richer, it is possible to argue that profit rates are high by an international measure — although not high enough to compensate for the low productivity of the rest of the economy. This, of course, is the crux of the matter.

The underdeveloped countries producing copper want a price which while it will not allow workers and technicians in this industry to enjoy a standard of living in the vicinity of their co-workers in the industrial countries — will enable diverse functionaries, officials, and the like *outside* the industry to maintain themselves in a fashion completely incommensurate with the contribution they are prepared to make to the economies of their respective countries. Under the circumstances it can be submitted that no price, regardless of how high it is, would suffice. It can also be submitted that this situation prevails in most under-developed countries, and is relevant for the entire range of primary commodities.

At the other end of the scale is the question of how far down the price of copper sold by the less developed countries could be reduced. This is for the most part an academic matter, with only a limited application to the real world. The first and most important reason for this is that if the price of copper were forced down by 25 or 30 percent from its present level, it would save the consuming countries a paltry 600 or 700 million dollars, but at the same time would net them the enmity of 40 or 50 million people in the producing countries. Obviously, this exchange is unacceptable. The second reason is that the technique for selling copper, which to a considerable extent involves long term contracts tied to the price prevailing on the London Metal Exchange, protects the seller from many of the advantages that the buyer has in this type of game. Were it simply a case of buyer-seller bargaining, without the intermediation of the metal exchange or a similar arrangement reflecting the actual supply-demand situation on the market, the price of copper could be pressed much lower than the 450 pounds mentioned above.[7] The logic here is as follows.

When a copper producer in a less developed country sells his copper he sells for pounds or dollars. His costs, however, are in both pounds or dollars (or a similar currency) *and* in his domestic currency. Where the first of these is concerned we have amortization and interest on machinery purchased abroad, payment for such current inputs as oil, and a few others of this type. For the second the largest entry is for salaries and wages. Take as an example a case where 250 pounds must be used to pay various debts abroad that are related to producing a ton of copper. This can be regarded as the first component of the price of copper. Now let us look at wage and salary costs. What we usually have is a situation where a miner might be, in local currency, among the highest paid employees in his country, but still does not enjoy the living standard of his colleague in the US or Canada. The reason for this is that a large part of such a standard would have to be imported, and the exchange rate relating the money in which he is paid and the foreign currency required to purchase these imports is usually such as to deny him access to these goods, or for that matter domestic goods having a large import content. In terms of the numerical example begun above let us say that, given the exchange rate, his entire wage in local currency is only sufficient to obtain him 50 pounds in foreign currency. This then would be the *maximum* value of the second component of the price: the labour cost in foreign currency.

There is one more component. This is the part going to the owner of the producing capacity, which in the present situation is usually the government of the country in which the capacity is located. The thing to appreciate about this component is that in a buyer-seller bargaining situation, the size of this component would be negotiable. My own observations of the world copper industry lead me to believe that up to now most sellers of copper are quite happy to rely on a system of pricing that obviates this kind of bargaining, since given the fact that most of these countries must either sell their copper or face bankrupcy, they might easily find themselves in a situation where this third component would be very small indeed.

It is of course always possible that attitudes will change, or mistakes will be made, and therefore it is not entirely unthinkable that a copper cartel will be formed that discards the present system of pricing. Assume that this does happen, and the price of copper is arbitrarily set at two or three times the present level. The way that the consuming countries can handle this is simply to accumulate an inventory of copper that could last 6 months or a year.[8] Most of the governments of the copper producing countries cannot wait 6 months or a year for income however, and eventually this price would have to be negotiated. As far as I can tell, all the advantages in this type of negotiations would be on the side of the consumers.

In case the reader does not believe that the advantages would be on the side of the buyers, he should remember that the buyers have other strategies to which they can resort. Perhaps the safest way to avoid the possibility of having to face a producer cartel of the type indicated above is to reduce the demand for copper. Two ways of doing this are to increase the substitution of other materials for copper — in particular aluminium; and to increase the amount of copper being won from scrap, or recycled. Where substitution is concerned, I like to think that it is possible to discern a kind of *optimal* pattern of substitution. If, for instance, the rate of substitution of aluminium for copper is compared for the major industrial countries, it will be seen that substitution has gone much faster in some uses in some countries than in others. (By uses here it is meant electrical wiring, construction, machinery, etc.) The optimal substitution pattern is then the pattern which has *each* country substituting for copper at a rate equal to the most rapid rate of substitution in any of the other countries.

If the governments of the industrial countries find the price of copper rising too fast because of what they regard as unethical practices by producers, they can resort to encouraging, by laws if necessary, consumers of copper to adopt the optimal pattern of substitution. For instance, if laws were adopted in all industrial countries prohibiting the importing of copper for purposes for which aluminium would suffice, the copper producers would find themselves in a very serious situation. On this point it should be mentioned that the copper industry has always displayed a high degree of sensitivity to the substitution problem, since history seems to indicate that once markets for copper are lost to aluminium, they are lost forever.

As for recycling, I have made a rough calculation indicating that with a rate of growth of copper use of about 3 percent per year, and an average length of life of products containing copper of 20 years, approximately 50 percent of the copper being used in a country can have its origin in recycling. This is a long run, or steady state arrangement, and for short periods could be exceeded because of the large volume of scrappable items now on hand in the industrial countries. (For instance, in the US there are about 20 million discarded automobiles littering roadsides and junkyards in which an intensive recycling campaign would be interested.) At the present time the amount of copper obtained through scrap amounts to about 40 per cent of total consumption, and an increase to 50 percent would put a not insignificant downward pressure on the price of copper.

Of course, the best arrangement for all parties would probably be an abandonment of the concept of economic warfare or confrontation between producers and consumers of copper. The price of copper could then be arranged to give the producers a satisfactory profit; and such things as recycling could still become more widespread because, given the finiteness of natural resources, a maximum amount of recycling makes sense for everybody in the long run. Substitution could also take place, but more slowly, since one of the driving forces behind substitution is the fear of copper users that copper prices might someday explode. Most important, it should be accepted that any counter measures adopted would be of a defensive nature, because unless the price was moved up by several hundred percent, and this took place in concert with similar price movements for other commodities, the economies of the industrial countries would hardly be disturbed.

Aluminium

Aluminium is produced by first crushing and drying bauxite ore, from which alumina is extracted. The alumina can then be smelted to produce aluminium metal. Bauxite, and a small amount of alumina, is for the most part produced in less developed countries; while aluminium is produced in the industrial countries. Among the industrial countries only Australia is a leading producer of bauxite; and France is the only OECD country that produces an important part of its consumption.

Bauxite is also a commodity that might, in the near future, come in for considerable attention, since some of the countries producing this ore have begun to discuss forming an OPEC-like cartel. Moreover, Jamaica has already, unilaterally, adjusted the price of bauxite upwards by a factor of almost four. This price rise caused no great stir, since it increased the price of aluminium to the final consumer by only 5 or 6 percent. The reason for this is that while, in a technical sense, bauxite is an indispensible input for most (but not all) of the aluminium being produced in the world right now, it does not amount to much in value terms. From the point of view of value, the most important input is energy.

Over the past twenty or so years, the price of aluminium has been on a kind of plateau. The price of oil has also been almost stationary, but not for the same reason. In the case of oil the immobility of the price can be understood in terms of the intention of the large oil companies to discourage investments in alternative sources of energy. Where aluminium is concerned the rate of expansion of the consumption of aluminium has been more than matched by the discovery of new reserves of bauxite. At the present rate of consumption there are reserves of bauxite sufficient to last almost 300 years. However given the present growth rate in the consumption of bauxite, these same reserves would last only slightly more than 50 years.

The problem that would face a potential cartel of the OPEC type among the less developed of the bauxite producing countries are many and varied. In the first place, most of these countries would seem to have little or no economic future outside of bauxite production, or the processing of bauxite. This rather unpleasant fact of life is well understood by them, and in fact the Jamaican price rise was not, apparently, intended as an example to be followed by the rest of the bauxite producers, but part of a programme to set the Jamaican economy in order. Another big stumbling block for a bauxite cartel would be the possibility of producing aluminium from material other than bauxite. The technologies for doing this would appear to have reached pilot form, and several of them — the Toth process and the Alcoa New Smelting process — claim to show sizeable cost reductions over present processes. According to the US Bureau of Mines these technologies should be operational in about a decade, at which time the possibility of the industrial countries being caught in some sort of bauxite shortages on the order of the oil shortage will rapidly decrease.

Although a number of individuals may find it in their interest to whip up interest in a bauxite cartel, the governments of the bauxite producing countries appear to be fairly careful in this matter. In particular they have not been convinced that in the long run conflict with the industrial countries would be in their interest, since where the new processes for producing aluminium are concerned — both those using bauxite and the other — these would seem to show a much higher efficiency in the use of electricity. Thus, even if bauxite prices did increase, these price rises would be offset by the decreasing cost of energy. This is a kind of ideal situation, since it would allow bauxite producers a higher price without forcing the pace of introduction of the new technologies — something that would have to take place in the industrial countries came to believe that bauxite was on its way to becoming scarce. It should be remembered here that aluminium is reckoned with as a substitute for copper, which means that it plays an important part in regulating the price of copper. A shortage of bauxite would not only cause an increase in the price of aluminium, but also tend to press up the price of copper, at least in the long run.

There is also another important issue here. The less developed countries that produce bauxite claim that they cannot get enough income from bauxite alone. It is partially their resentment of this situation that leads them to toy with the idea of forming an OPEC type cartel, even though in the present situation they have none of the advantages of the oil producers. One of the things they desire is to have more alumina producing, and aluminium smelting capacity in their countries. An argument for this is that income gained from bauxite is small as compared to that gained from further processing of this ore. An argument against is that in the production of aluminium there are important economies of scale, and these could not possibly be realised by installations located in the less developed countries.

But more important, there is the question of energy. Australia, for instance, is a rich country with a very large bauxite production, but it still pays them to ship this bauxite to Japan for further processing rather than to invest in the energy producing facilities that would allow them to do it domestically.

Still, this is one of those cases in which the economic argument against cannot be taken at face value. As it stands now, if the price of bauxite were to rise sharply, or if there was a halt in bauxite production, it would decrease the profitability of alumina and aluminium producing plants in the US and Europe. On the other hand, if there were very large investments in alumina and aluminium production, as well as energy supply, in the producing countries, then they would have a vested interest in keeping this capacity in operation.

There is, however, another argument for the location of more processing capacity in the primary producing countries. Today the industrial countries supply the Third World with hundreds of millions of dollars of financial "aid" that is wasted, either on the giving or receiving end. Rather than prolong the absurd pipedream that this type of aid contributes to the development of the receiving countries — a pipedream authored by bureaucrats in the industrial countries and the UN who have managed to make excellent careers supervising this aid — it might be best to earmark it directly for industrial development in these countries. Given the lack of productivity of other sectors, this is almost certain to mean an extension of the processing of primary production.

This paper has attempted to treat some aspects of the controversy between the producers of raw materials in the less developed countries, and consumers in the industrial countries. The two commodities examined, copper and aluminium, are indisputably the most important among the non-fuel minerals — at least where international trade is concerned — and involve most of the problems that would have to be taken up if any of the others were considered.

No attempt was made to look at deep sea mining, or such things as extracting minerals from seawater, mining at deeper levels, etc. Nor, for that matter, were any of the many non-economic issues relevant to this topic covered in depth. But on this score one more thing should be added. The underdeveloped countries will have to find their own way in this matter of minerals and development, but regrettably this is going to involve more than attending conferences in Geneva or Vienna. This is not going to make things easier for them, because the idea of affluence without effort has a strong hold on the elites in many less developed countries, and I cannot see this changing in the near future.

As for the industrial countries, the long run answer to the problem of mineral supply is probably going to involve an extensive alteration of consumption patterns, with sport and culture — to include education — playing a much larger part in the scheme of things. At least, this is my answer, and brushing questions of modesty aside for the moment, I have yet to hear one that makes more sense.

Notes and References

[1] In billions of dollars for nine major minerals: Bauxite, Copper, Iron Ore, Lead, Manganese Ore, Nickel, Phosphorus, Tin, and Zinc.

[2] Estimated.

[3] This problem is gradually accumulating a fairly rich literature, beginning with: Hotelling, H "The Economics of Exhaustible Resources", *Journal of Political Economy*, April 1931. Pages 137—75.

[4] In the case of buildings, for example, investments were made in heating plant that should have been made in insulation.

[5] If we make our calculation using 1973 as the base year, the *real* price of oil initially rose to 12 dollars. As for adjustments, the problem is that these were the wrong kind — namely, inflation and unemployment.

[6] See, for example, Banks, F E *"The World Copper Market: An Economic Analysis"*, Ballinger Publishing Company, Boston, 1974; and "Copper is not Oil", *New Scientist*, 1 August 1974.

[7] Although a marginal market in the sense that only a very small percentage of the copper sold in the world is physically traded on its premises, the London Metal Exchange (and in the US Commodities Exchange) reflects very closely the supply-demand situation in the world for a number of commodities.

[8] Remember that this stock would only amount to about 40 percent of the yearly consumption of copper, since this is the amount supplied at present by the less developed countries. The cost of a stock that would last one year would be about 1.1 billion dollars plus storage costs. This is no small amount, but the reader should remember that it would be spread across some of the richest countries in the world, and that at present the non-socialist industrial countries are carrying stocks of oil that cost at least twenty times as much. In addition, when the cartel broke down, the carrying costs, and perhaps some of the capital costs, could be shifted to the producers in one sense or another.

RESOURCES, PRODUCER POWER AND INTERNATIONAL RELATIONS

Robert Dickson and Paul Rogers

Introduction

The underlying theme of this book is that the most significant long term effects of trends in world resource use will be on the development prospects of the less developed countries and on the relationships between these countries and the developed industrialised nations of the world. This is at variance with much that has been written since the onset of the environmental debate in the late 1960s. Consideration of the future availability of important resources has been very largely limited to prospects for the major resource consumers, the industrialised countries, and little attention has been paid to the total world scene in general and the prospects for less developed countries in particular.

Even in the current era of increasing awareness of "producer power", the effects of such a development continue to be considered almost solely in terms of future availability of resources for the industrialised nations. This is indeed a criticism that can be levelled fairly at the whole of the environmental debate extending to the work and outcome of the UN Conference on the Human Environment in Stockholm in June 1972. Too often, attention has been concentrated on the limited, almost parochial, questions of the environment and resource problems of the rich nations, and few commentators have seemed aware that action by major resource consumers among the less developed countries could have a fundamental effect on the future economic prospects of the industrialised nations.

The aim of this paper will be to show that there is in progress a fundamental change in the perceptions and realities underlying the relationships between industrialised and less developed nations, that this constitutes the most significant change in international relationships in the last few decades, and that it has the most profound implications for world development and *therefore* international stability in the closing decades of the present century.

We say "therefore" because the two are unlikely to be compatible. The stability of the present international economic order is predicated upon the continuance of the situation in which the "rich north" has overwhelming political and economic power, and uses it so to structure international trading relationships that this power, and ensuing wealth, is maintained at the expense of the weakness and poverty of the "poor south". International stability may be affected because historical evidence suggests that the rich rarely concede their wealth to the poor except in response to force.

The 1950s and 1960s were, for the industrialised nations, decades of unparalleled economic growth. Along with this went increasing rates of consumption of raw materials. The standard of living for the great majority, but not all, of their people increased greatly and the so-called "consumer revolution" ensured that most people in these countries came to assume that increases in wealth would be endless. Of the many factors which gave rise to these "good years", two stand out as being particularly important. One, undoubtedly, was the application of a wide range of major technological innovations of the 2nd World War to industrial life. This is true not just in the field of mass production techniques and improvements in transportation, but equally so with developments in communications, petrochemicals, medicine and energy utilisation. The jet engine, computers, antibiotics, plastics and atomic power all owe much of their origins to war-time conditions and their development did much to bring about that consumer revolution which the richer nations of the world now accept as their natural birthright.

As important as these technological developments was the fact that the nations which controlled these also had access to plentiful supplies of energy and raw materials, not just within their boundaries, but from among the many less developed countries of the world. Even after the effects of the independence movements of the post-war years, the real control of resources in many apparently independant ex-colonies lay with multi-national corporations whose allegiances, in turn, lay with the industrialised countries.

Not only this but the less developed countries, as producers of primary commodities or raw materials, consistently found themselves unable to obtain the price for their resources to match accelerating prices for manufactured goods which they could neither produce for themselves nor export to industrialised countries.

It is a fact of international trade that, during the 1950s and 1960s especially, the less developed countries were in a trade trap, unable to earn, from exporting their raw materials, the foreign exchange necessary to ensure their own development. For a short time, the 1973/4 boom in primary commodity prices partially rectified this situation, but it was short-lived and did little to recompense the producer countries for years of decline.

Only with the advent of producer power in 1974 did the power base of world trade start to shift markedly away from the industrialised nations. Now, few people will deny that less developed countries have at least a nascent power and our concern here must be with how they exercise that power, and equally important, how the industrialised nations seek to respond to this totally new situation.

If we are to make any progress in attempting such an exercise in long term forecasting, we must first examine the recent history of trading relationships between industrialised and less developed nations, and also the origins of the realisation by leaders of less developed countries that they are no longer so powerless in seeking a redistribution of the use of the world's resources.

Elsewhere in this book, will be found a detailed consideration of trading patterns. To complement this the initial intention here is to discuss relations between industrialised and less developed countries as expressed in the work of the UN Conference on Trade and Development, (UNCTAD) without doubt the most important single forum for negotiations between rich and poor countries in recent years. We will then examine the origins and work of the 1972 UN Human Environment Conference and the manner in which it promoted the theme of a planet with limited potential for resource use. We will then consider trends in the exercising of producer power in the crucial first eighteen months following the increase in oil prices achieved by the Organisation of Petroleum Exporting Countries later in 1973.

Finally we will make some suggestions concerning possible future developments in this field and their likely effects on international stability.

The History and Development of UNCTAD

The United Nations Conference on Trade and Development has, at the time of writing, held three major sessions, UNCTAD 1 in Geneva in 1964, UNCTAD 2 in Delhi in 1968 and UNCTAD 3 in Santiago, Chile in 1972. A fourth session is planned for Nairobi in May 1976. These four-yearly meetings have taken the form of major bargaining sessions between industrialised and less developed nations and their results show clearly the nature of the political and economic relationships between these groups of countries.

Towards the end of the Second World War and in the years immediately after it, the United Nations Organisation and a series of associated international specialised agencies were established. These included the World Health Organisation (WHO), the Food and Agriculture Organisation (FAO) and the United Nations Education, Scientific and Cultural

Organisation (UNESCO). A number of governments instrumental in the formation of the United Nations Organisation were interested in seeing established, in addition to those just mentioned, a trio of specialised agencies to be responsible for restoring and maintaining international economic order in the post war years. These were to be the International Monetary Fund (IMF) concerned with the world's monetary system, the International Bank for Reconstruction and Development (World Bank), concerned with financing reconstruction and development on a multilateral but nevertheless commercial basis, and finally an International Trading Organisation.

The years leading to the Second World War had seen a proliferation of restrictive and restricted trade agreements on all manner of trade items between many groups of countries. The main aim of the planned ITO was to work for a liberalisation of the world trading system by the ending of such barriers to free trade.

ITO itself was never established, the charter being unacceptable to a number of member governments of UNO, but a multilateral trade treaty was established, arising phoenix-like from the ashes of ITO, as a result of a tariff negotiating conference in Geneva in 1947. This was the General Agreement on Tariffs and Trade, administered by the surviving remnant of ITO the Interim Committee for the International Trading Organisation[1].

But the GATT, as its name implies, was little more than an international agreement between nations administered by a relatively small secretariat. Its aim was to work for a free trade situation in world markets, this being considered as a means of ensuring full employment and economic development in participating countries. Thus, according to the Preamble to the Agreement, the signatories would enter into

"reciprocal and mutually advantageous arrangements directed to the substantial reduction of tariff and other barriers to trade and to the elimination of discriminatory treatment in international commerce"

so that international trade relations were conducted with a view to

"ensuring full employment and a large and steadily growing volume of real income and effective demand, developing the full use of resources of the world and expanding the production and exchange of goods."

In practice, the GATT has consisted largely of a series of tariff bargaining sessions when UN member nations have met in talks aimed at liberalising trade, with probably the most famous talks being the 1964—7 sessions in Geneva known as the Kennedy Round.

It should be clear from this brief account, that the GATT was set up as an instrument for liberalising world trade. It was certainly not implicitly concerned with the problems of the less developed countries and in spite of an expansion of work in this field undertaken by the GATT secretariat, the GATT as a whole came under increasingly heavy criticism in the 1960s from many less developed countries for failing to be the vehicle they required to give them a stronger position in their trading relationships with industrialised countries.

Indeed the whole idea of a progressive liberalisation of world trade was becoming, in the view of governments of less developed countries, an unacceptable idea. Their belief was that, in practice, free trade was anything but free except for the rich partner, free trade seemed to mean a freedom for the rich to get richer at the expense of the poor. This certainly does not do justice to the genuine aim of many people connected with the GATT's work on the trade problems of the world's poor, but nevertheless the fact remains that towards the end of the 1950s, less developed countries were coming increasingly to the view that a positive restructuring of world trade in their favour was required to rectify a world trading structure weighted heavily against them. To them, the GATT and its seeking of free trade was no more than a stepping stone to a fairer system of world trade.

It became clear that such a move was well beyond the scope of the GATT, dominated as it was by the industrialised countries, and third world governments began to seek a new forum within the United Nations Organisation in which such trading developments might be discussed and implemented. These moves were resisted by many western countries, who favoured the continuation of the GATT as the major UN body concerned with trade. An additional factor in their support for the GATT involved a cold war dimension. The Soviet bloc did not support the GATT, finding it structured in such a way as to make it somewhat deficient in dealing with the trade problems of centrally planned countries. In the cold war climate this provided an added reason for western support of the GATT.

In the event, it was pressure from the increasing number of newly independant states present within the United Nations which finally lead to the creation of UNCTAD. By the early 1960s, they were able to succeed in their demands, UNCTAD was established with its first session fixed for 1964 to be preceeded by at least two years of preliminary work. The appointment, early in 1963 of the Argentinian economist and former Secretary-General of the UN Economic Commission for Latin America, Dr Raul Prebisch, was a major step towards giving UNCTAD an independant voice and this was in evidence by the end of 1963 with the publication of a report entitled "Towards a New Trade Policy for Development" or more commonly the "Prebisch Report". This document, prepared by the new UNCTAD secretariat but owing something to a position paper presented by a group of 75 less developed countries at the autumn 1963 session of the UN General Assembly, essentially gave the bargaining position of the less developed countries at Geneva in 1964.

The analysis of the trade and development problems of less developed countries which was contained in the Prebisch Report was criticised at the time, especially by economic advisers in industrialised countries. But it gained support from less developed countries in seeking a number of changes in international trade and aid designed to rectify the trading disadvantages experienced by them. Briefly the major proposals of the report may be summarised as follows:

1. The raising of prices of primary commodities produced by less developed countries, this to be done primarily through multilateral commodity agreements.

2. A "compensatory finance system", funded by industrialised countries which would be designed to compensate less developed countries for any deterioration in terms of trade persisting after the introduction of commodity agreements.

3. A tariff preference system whereby industrialised countries would give preferential treatment to manufactured and otherwise processed goods from less developed countries.

4. Schemes to encourage regional industrialisation among neighbouring less developed countries thus achieving economies of scale.

5. Increased participation by less developed countries in shipping and freight insurance, at that time activities which were almost entirely in the hands of a few industrialised countries such as Britain.

6. The softening of aid to reduce existing debt burdens and improve the quality of future aid.

Three comments may be made concerning the Prebisch Report. One is that the proposals added up to a massive change in the world trading system which might alter the balance of power away from the industrialised countries. Few observers had any illusions that such proposals would be acceptable to the governments of the major industrialised nations; they constituted an initial but basic bargaining position for the less developed countries.

Secondly, it provides an excellent example of the UN conference mentality which has been so prevalent since the inception of the UNO. It cannot be stressed too strongly that, where political realities are concerned, UN conferences may mirror or even clarify such realities but can do very little about changing them, — that requires a coercive power which the UNO does not possess.

A third point, with the benefit of hindsight, is that, in the present era where power may be passing to the less developed countries through their exercising of producer power, an updated version of the points made in the Prebisch Report, far from being unrealistic, might well be the minimum demands to be made at future UNCTAD and other UN meetings.

But to return to the first session of UNCTAD, the March to June meeting in Geneva in 1964, it was apparent at the start of the Conference that many delegations from industrialised nations viewed the proceedings with some suspicion, and many of the scores of less developed countries represented had varying policies towards the Prebisch Report. In the event, UNCTAD 1 saw the loose alignment of several regional groupings of less developed countries created, with an overall group known as the Group of 77 playing a major role in coordinating activities by the delegations of the less developed countries.

It became apparent that these countries had a numerical voting superiority but this was meaningless as motions would not be enforced, nor would their substance be voluntarily conceded. It was therefore felt to be far more important to negotiate agreements which had the support of all parties. Given the divergent interests of the parties concerned, it was hardly surprising that, in this respect, the achievements of UNCTAD 1 were minimal.

Very few of the Prebisch proposals made any headway, those that did being concerned with supplementary financing and tariff preferences. In the first case, the World Bank was asked to make a feasibility study to be completed by the next full meeting of UNCTAD and in the second case, tariff preferences, a general motion accepting the need for such preferences was agreed in principle, and the UN was asked to set up a group to translate this into practical formulae.

Thus the achievements of UNCTAD 1 were minimal, its main value probably lying in the way in which it enabled many less developed nations to forge links with each other. Over the next four years, terms of trade for many less developed countries declined still further, the newly acquired independance of several former colonies increased the Group of 77 to ninety, and indications from the OECD and the United States suggested that UNCTAD 2 in New Delhi in February and March 1968 might be a more fruitful meeting than UNCTAD 1 had been. Supplementary finance and a general scheme of preferences were two of the main subjects for consideration, together with the need for more commodity agreements and the necessity of greatly increasing the quality and quantity of development assistance to less developed countries from industrialised countries.

In the event, UNCTAD 2 was, from the point of view of delegations of "less developed countries, hardly a success." There was no general agreement on either tariff preferences or supplementary financing. The creation of further commodity agreements made virtually no progress and although the industrialised nations confirmed their belief in the need to raise aid levels to 1% of GNP, no date was fixed for achieving this.

Part, at least, of the failure of UNCTAD 2 was due to a lack of unity on the part of the less developed countries, most apparent in discussions on the proposed generalised scheme of preferences, but unity was only required in the face of the attitudes of the industrialised nations, where the opinion was growing that the UNCTAD sessions were annoying occasions when delegates of less developed countries indulged in rhetoric and put forward policies well out of line with the interests of the industrialised nations.

In view of this it may seem surprising that the organisation survived to its third session in Chile in 1972. In part this was because the discussion of preferences within UNCTAD was instrumental in the establishment of a number of bilateral and multilateral preference schemes, and UNCTAD also tended to stimulate further attempts to negotiate commodity agreements. Particular emphasis in UNCTAD 3 came to be placed on the preferences negotiations, many less developed countries being hopeful that individual agreements could be expanded into a general scheme.

But the Santiago session involved a great deal more than preferences. In addition to the usual questions of increased aid, and expanded commodity agreements, further issues involved possible reforms of the international monetary system, a linkage between the IMF's Special Drawing Rights Scheme and development assistance, and an improvement in the stake of less developed countries in invisible earnings in such areas as tourism, shipping and insurance.

UNCTAD 3 took place shortly after the end of the first UN Development Decade (1961–70) with a world development background which was hardly auspicious. Most observers accepted that the well-being of many of the inhabitants of less developed countries had probably not improved during the decade, and that prospects for the Second Development Decade (1971–80) did not look hopeful. Against this background it could be argued that UNCTAD 3 provided one last chance for a greatly increased degree of co-operation between industrialised and less developed nations to ensure a more just world economic order.

In the event, the third session of UNCTAD, held during April and May 1974 in Santiago, was disastrously unsuccessful, even more so than the New Delhi session four years previously. No further agreement was reached on the general system of preferences, and no new commitments were made on aid. Commodity agreements made virtually no progress, except, perhaps, in the case of cocoa. Less developed countries marginally increased their strength on the International Monetary Fund, but the SDR link made no progress and even resolutions on invisible earnings were watered down. As one commentator noted, the meaning of UNCTAD seemed to have become "Under No Circumstances Take Any Decisions".

UNCTAD 3 and Producer Power

The effect of the outcome of the UNCTAD 3 session is highly relevant in the context of producer power. Only 15 months before the first major example of producer power, the October 1974 OPEC oil price rises, the governments of less developed countries were faced with the realities of the current world economic order, namely that the industrialised countries were not prepared to accept any changes in that order which might on the one hand adversely affect their standard of living while on the other, greatly improve the development prospects of the less developed countries. Essentially the latter were on their own and a remarkably prescient leader in the London newspaper "The Guardian" discussed the anger which these countries felt after UNCTAD 3[2].

In addition to making industrialised countries aware of the feelings of the delegations from less developed countries, the leader went on to say that:

"the anger has also served to produce rather more cohesion among the third world countries themselves. The Francophone African States have in the past been accused of slavish adherence to European interests; this time they were willing to vote solidly with the Group of 77, just as the Ivory Coast has at last been prepared to join with Brazil to stockpile coffee and force up its price, rather than wait any longer for the broken-backed International Coffee Agreement to deliver results. Producers of oil and coffee, the two biggest third world exports, are now exploiting their market power, and these tactics may spread. This is no cure-all — too many tropical products face synthetic competition, and some key minerals are found abundantly in the rich countries — but this kind of solidarity should certainly improve the bargaining power of the third world".

Thus UNCTAD 3 was the latest in a series of confrontations which served only to confirm the unwillingness of industrialised nations to commit themselves to world development.

The Stockholm Environment Conference

If an appreciation of the need for producer power was the legacy of UNCTAD 3, then that notion received a powerful boost from a somewhat unexpected quarter, the UN Conference on the Human Environment held in Stockholm in June 1972. The Stockholm environment conference represented the summit of a period of increasing interest in environmental affairs in many industrialised countries, an interest which dated back over a decade to worries about environmental contamination with pesticides and other pollutants.

The original impetus for the conference had come from Sweden, which, back in 1968 had called on the UN

"To provide a framework for comprehensive consideration within the United Nations of the problems of the human environment in order to focus the attention of governments and public opinion on the importance and urgency of this question"[3].

The main aim of the conference was therefore one of opinion-forming, but this was later amended to give it more of an "action-planning" orientation.

The conference took about two years to plan and ended up as an excessively bureaucratic affair, with its main emphasis on the environmental problems arising from current or past industrial activity. Other problems were considered, but a major criticism of the conference was its excessive concern with the relatively limited environmental problems of the developed industrialised nations. Even those world-wide problems which were scrutinised were mainly those likely to affect the industrialised nations such as marine and atmospheric pollution.

Although heavily weighted towards the problems of the industrialised nations, preliminary conference documents did something to assuage the fears of governments of less developed countries that their problems would not be considered, and some emphasis was placed on the need for an equitable sharing of the world's resources and on the need to promote the ecologically sound economic development of less developed countries. In practice, other factors were far more important in ensuring that participation of delegations of less developed countries was forthcoming. Chief among these were the efforts in this direction of the Canadian Secretary-General of the Conference, Maurice Strong, who was remarkably successful in convincing governments of less developed countries that the meeting would be relevant to their needs.

In the event, the real significance of the Stockholm conference lay not so much in the official resolutions as in the unofficial interest generated during the conference in the "Limits to Growth" debate. That well-known MIT study on world dynamics had suggested that, unless the then current trends towards increasing pollution, resource use and other expressions of human activity were controlled, there would be a progressive breakdown in world order.

In most parts of the world the book "Limits to Growth", a somewhat popularised account of MIT studies[4], was published only a few months before the Stockholm conference and such was the timing that its ideas were widely disseminated during the conference. This is not to say that the book was received with acclaim. It aroused considerable opposition from several camps. Among these were the proponents of economic growth, maintaining that society in general and technological developments in particular would be capable of finding solutions to environmental problems as different as atmospheric pollution and resource depletion. Perhaps more serious were the criticisms that the study did not adopt a realistic picture of the world economic order, largely ignoring the realities of wealth and poverty and considering future trends largely in terms of industrial society. Strong criticism was expressed that "Limits to Growth" was little more than an attempt to dissuade the less developed countries from following the paths of economic and industrial development which had so altered life in the industrialised countries but which would increase competition for resources.

Finally there were criticisms of a more technical nature concerning the validity of the modelling techniques involved in the study. But these and other criticisms did little to detract from the key message of Limits to Growth. In effect it was a retelling of the old biological law that you cannot have infinite growth in a finite system. The world, considered as an essentially finite system, thereby has a finite capacity for supporting human activity.

Opinions differed greatly on the question of maximum permissible levels of activity, with many environmentalists arguing that the world was rapidly approaching those limits. But one aspect of the question was recognised to have the most basic implications for the prospects of less developed countries. In the words of the British ecologist Palmer Newbould, in a paper prepared a few months after the Stockholm conference:

"My own belief is that however successful population policies are, the world population is likely to treble before it reaches stability. If the expectations of this increased population were for example, to emulate the present life style and resource use of the USA, the demand on world resources would be increased approximately 15-fold; pollution and other forms of environmental degradation might increase similarly and global ecological carrying capacity would then be seriously exceeded. There are therefore global constraints on development set by resources and environment and these will require a reduction in the per caput resource use of the developed nations to accompany the increased resource use of the developing nations, a levelling down as well as up. This conflict cannot be avoided. It became the central theme of the United Nations Conference on the Human Environment"[5].

It is no doubt true that powerful arguments can be generated to oppose that view, but the important point is that it was a view of world affairs totally opposed to the usual approach and it was a view which received a 'sympathetic hearing' from many delegates of less developed countries. Essentially it ran entirely counter to the traditional view of development studies; that the less developed countries would benefit and ensure their development primarily in parallel with the increasing wealth and economic development of the industrialised countries — "crumbs from the rich man's table" as it has been put.

But this new view meant that the less developed countries could only finally achieve a real measure of development at the expense of the industrialised countries. A re-allocation of world wealth and the consumption of the world's resources was required. Not only this, but the very debate which had led to the promotion of this view also showed a mechanism for achieving this transformation. In an era of increasing resource use and decreasing resource availability, the less developed countries, primarily the exporters of the world's raw materials, already had within their borders the sources of power to effect the change. Essentially, producer power, the possession and control of those previously cheap raw materials, was the one asset of the less developed countries. It may well be that for all its concern with the problems of the industrial nations, the Stockholm conference will go down in history as the forum which provided a major impetus for the notion of producer power. The less developed nations had that which the industrialised nations would increasingly require, physical and biological resources, and were becoming in theory if not yet in practice, countries with power.

The Summit Meeting of Non-Aligned Countries, Algiers 1973

Dr Dasgupta's paper elsewhere in this volume gives a detailed account of the development of OPEC, and shows that even by early 1973, that group of producer countries was beginning to exercise a commodity bargaining potential on a hitherto unparalleled scale. That development, together with the effects of the failure of UNCTAD 3, and the discussions at the Stockholm conference all contributed to an accelerating interest in the possibilities for producer power which was clearly reflected in speeches and resolutions at the Fourth Conference of Heads of State or Government of Non-Aligned Countries in Algiers in September 1973.

The non-aligned summit consisted primarily of a large number of less developed countries together with a few relatively neutral industrialised countries such as Yugoslavia. It was the fourth such meeting in a series going back to Belgrade in 1960 and owed much of its origins to the efforts of President Tito of Yugoslavia, having originally been formed largely to promote the dislike that many neutral countries had for cold-war politics and the threat of nuclear war. But the Algiers summit meeting was significant in the way in which these previously weak countries came together in a mood of growing awareness of their latent strength.

This was expressed succinctly by President Tito when, addressing the conference, he said that the resource bases of the non-aligned countries should be seen to constitute

> "a latent economic power of the non-aligned which, if utilised correctly, could greatly strengthen the position of non-aligned countries in economic exchanges in the contemporary world"[6].

It was a theme taken up by other conference participants, and it was also expressed in the Economic Declaration of the conference, the full text of which is included in Part II of this volume. The Declaration included a section on "Sovereignty and Natural Resources" which was adopted by the conference and embodied a comment on producer power:

"The Heads of State or Government recommend the establishment of effective solidarity organisations for the defence of raw material producing countries such as the Organisation of Petroleum Exporting Countries and the Inter-governmental Committee of Copper Exporting Countries (CIPEC), which are capable of undertaking wideranging activities in order to recover natural resources and ensure increasingly substantial export earnings and income in real terms, and to use these resources for development purposes and to raise the living standards of their peoples".

The Sixth Special Session of the UN General Assembly and the Dakar Conference

Within a few months of the Algiers meeting, the full effects of the success of the OPEC countries in exerting producer power were beginning to be felt. As a result of this, and of the short-lived commodity boom of 1973/4, a special session of the UN General Assembly was convened in April 1974 to discuss problems of raw material supplies. This was the first such session of the UN General Assembly to be concerned with economic affairs, an indication of the rapidly growing importance of producer power. An account of this session and the major resolutions are included in this volume and the main results to be drawn from the meeting were that the participants from less developed countries were already seeing commodity bargaining as a means of effecting a new international economic order, one which would greatly improve their development prospects.

Perhaps as significant as the special session was the meeting of less developed countries held in Dakar about a year later. This was a five day meeting in Senegal in February 1975 for raw material producers and was attended by delegates from 110 less developed countries together with just three industrialised countries present as observers, Austria, Finland and Sweden. The important outcome of the Dakar conference was that it showed that the financially powerful member states of OPEC might well be prepared to aid other much weaker nations in exercising producer power.

This development was highly significant. Only a few days before the start of the Dakar meeting, the UNCTAD Commodities Committee began its consideration of a massive plan to stabilise the prices of 18 primary commodities representing an estimated 60 per cent of world trade in raw materials excluding oil. The $11,000 million buffer stock arrangement to back up this plan was intended to be financed by both importing and exporting countries, but at the time of the Dakar conference it was clear that heavy opposition was likely from a number of industrialised countries. If this turned out to be the case, possibly at UNCTAD 4, then it became apparent at Dakar that the less developed countries would be prepared to go it alone. Without the financial backing of OPEC countries this would be impossible, but the indications were that some of the OPEC countries, with massive financial reserves, might well be prepared to provide just that backing; probably in return for support on political questions.

Producer Power in Action

Thus two major international meetings held within three years of UNCTAD 3 showed that the *potential* position of many less developed countries was changing dramatically, that they were aware of this change and that they were prepared to use their developing power. But to what extent had concrete examples of producer power for commodities other than oil occured during this time?

It is here evident that only rock phosphate production has been anywhere near as successful as oil, but there were numerous examples of moves towards producer power, even though those moves were no more than preliminaries to longer term activities.

The example of rock phosphate is interesting in showing the way in which short term supply-demand changes can suddenly increase the power of the producer. In January 1974, Morocco doubled the price of its rock phosphate exports and in July of that year it increased the price by a further 50%. The background to this is that although Morocco held about 50% of the world's proved phosphate reserves and had a powerful position in world phosphate exports, it had not, prior to 1973, found itself in a position to dictate prices.

The changed circumstances were that 1973 saw demand outstrip supply by around 5% so that Morocco suddenly found itself able to exert producer power. The argument was that these price increases were doing no more than redressing a declining trading situation which had existed for decades. Thus the actual price of rock phosphate exports had actually declined over a 20 year period, the decline in real terms, allowing for inflation, being considerable.[7]

It has been made clear elsewhere (eg. Dr Bank's paper) that major difficulties face efforts by copper and bauxite producers to exercise producer power and by the time of the Dakar meeting they had not succeeded to any extent. Yet an International Bauxite Association had been established and even the disunited CIPEC had actually undertaken action to try and bolster copper prices by 15% cuts in output. The Association of Natural Rubber Producers was becoming more active, an association of iron ore producers was in process of formation, there was evidence of increased activity by coffee and banana producers and even relatively unimportant commodities like mercury, tungsten and pepper were becoming subject to preliminary discussions among the producers.

Thus the 18 month period from the onset of the massive oil price rises in September 1973 to the Dakar conference in February 1975 marked a turnabout in the attitudes of the governments of less developed countries when compared with the situation at UNCTAD 3. The results may have been minimal apart from oil and phosphates, but the political will appeared to be emerging.

Prospects for Producer Power

In summary there is now plentiful evidence to support the view that a large number of less developed countries are seeking to use their primary commodity production as a means to improve their development prospects. Their intention is to exercise producer power. But what are their prospects for success, remembering that they seek to gain economic advantage from nations which are, in most respects, far more powerful than them, individually or collectively?

Let us first consider possible scenarios. Early in 1973 Dr Edwin Brooks put forward three possibilities and his views deserve an extended quote:

"Three alternative scenarios can be postulated. First that the rich world will be stripped of its privileges and wealth by the bargaining strength of the developing countries. This was the coercive process exemplified during the OPEC negotiations; but there can be no certainty that such favourable circumstances would generally operate to the benefit of other third world producers.

"However, coercion can take on many guises, and with nuclear weapons already extending throughout Asia we cannot rule out the possibility of local conventional wars against the rich world, and its overseas investments, in which the latter's nuclear amoury could not in practice be used to offset the weight of numbers. The lesson of Vietnam is that as long as the Third World country has powerful nuclear partners in the wings, it can take on and defeat a rich super-power. With China, and perhaps India too, likely to possess a massive nuclear stockpile and ample ICBMs by the nineteen-eighties the scope for limited wars under the nuclear umbrella is unlikely to diminish as global tensions, envy and frustration accumulate.

"The second scenario also postulates coercion and intermittent military engagements, but this time the picture is reversed, with the rich countries using their power nakedly to retain their privileges. There is a hint of some such trend in the tacit understanding between the two major super-powers about their respective spheres of influence;

thus the Red army is "permitted" to invade Czechoslovakia with tanks, while the USA is "permitted" to bomb North Vietnam with napalm. Admittedly such examples of *real politik* (to which might be added in the post-war era Guatemala, Cuba and the Dominican Republic) serve to show how precarious are the conventions underlying the balance of terror, yet the common interest of the nuclear powers in seeing that their nuclear powers are never used might well develop into a tacit agreement — which even China might come to endorse — that troublesome and shrill countries of the Third World should not have delusions of affluence, run amok, and upset the apple cart of global peace and security.

"The third hypothetical scenario is presumably the one which the civilised liberals among us would prefer. That is, a world in which the rich countries accept that in their own long-term interests (including those of peace and security) it is vital to avoid a crowded glowering planet of massive inequalities of wealth buttressed by stark force and endlessly threatened by desperate men in the global ghettoes of the underprivileged."[8]

There is certainly a danger of categorising possible future trends into discrete scenarios, but they serve as a valuable basis for discussion and we may add a further scenario to those of Brooks. This would be a much more overt confrontation between less developed and industrialised countries leading to a stalemate with each side the loser.

But what of the three scenarios proposed? How do the prospects appear for the remaining quarter of this century? We will take the scenarios in reverse order, starting with the hoped-for co-operation between rich and poor nations.

Prospects for Co-operation

Two separate considerations suggest that this is unlikely. The most important is the obvious fact that it would represent an almost unbelievable change in attitude by the developed industrialised nations, all the more so in that it would involve a voluntary forgoing by them of some measure of their wealth. We paid considerable attention, earlier in this paper, to the history of the UNCTAD negotiations. The lesson to be learnt from the UNCTAD sessions, from the results of the first development decade and from the current state of so many third world countries is that the industrialised nations are not prepared to alter the world economic order in any manner likely to improve the development prospects of the less developed countries, as this would adversely affect their standard of living. Historical evidence suggests that the rich rarely concede their wealth to the poor except in response to force.

However, more recent and particular evidence suggests that the industrialised nations will make little more than gestures towards producer power, — attempts to "buy off" the less developed nations and thereby counteract tendencies towards producer power. This came out most clearly in the proceedings of the meeting of Heads of Commonwealth Countries in Jamaica in April 1975.

On that occasion, the British Prime Minister, Mr Harold Wilson presented an outline plan for improved commodity marketing arrangements designed to stabilise commodity markets, a plan given in full in Part 2 of this book. Also included is the statement of reply by the Prime Minister of Guyana, Mr Forbes Burnham. In essence, Mr Wilson sought agreement to establish commodity plans to ensure stable prices and adequate supplies for consumers, the two working together for the benefit of producers and consumers. As a basis for discussion he suggested six general commitments from member governments:

"First we should recognise the interdependence of producers and consumers and the desirability of conducting trade in commodities in accordance with equitable arrangements worked out in agreement between producers and consumers.

Second, producer countries should undertake to maintain adequate and secure supplies to consumer countries.

Third, consumer countries for their part should undertake to improve access to markets for those items of primary production of interest to developing producers.

Fourth, the principle should be established that commodity prices should be equitable to consumers and remunerative to efficient producers and at a level which would encourage long term equilibrium between production and consumption.

Fifth, we should recognise in particular the need to expand the total production of essential foodstuffs.

Sixth, we should aim to encourage the efficient development, production and marketing of commodities (both mineral and agricultural) — and I should like to emphasise forest products — and the diversification and efficient processing of commodities in developing countries. We should not deduce from two centuries of history that there was any divine ordinance at the creation of the world under which providence deposited the means to primary production in certain countries and it was ordained that those minerals, or other products should be exclusively or mainly processed in other countries."

Throughout his statement, Mr Wilson laid emphasis on the good management of commodity markets for the benefits of producer and consumer alike. In many ways it was the kind of statement which would have been welcomed at a meeting such as UNCTAD 1, 2 or even 3. But the results in Jamaica were rather different. The conference agreed to set up an expert group to consider the matters but failed to indicate any corporate welcome or rejection of the Wilson proposals.

Individually, though, leaders of less developed countries within the Commonwealth made it clear that a package of commodity proposals would not be sufficient to produce the new international economic order that was discussed at such length in the sixth special session of the UN General Assembly the previous year. This is clear from the statement of the Guyanan Prime Minister Mr Forbes Burnham:

"I have listened with great interest to the proposals put forward by our good friend and colleague Harold Wilson. As he has rightly said "we cannot negotiate a general agreement on commodities here at Kingston." We cannot, of course — because that is a matter for the international community as a whole. Nor can we pre-empt the debate upon them by premature endorsement, or, conversely, by premature rejection. What I can say for our part is that we cannot but look askance at piecemeal proposals which confine themselves to merely one area of the Programme of Action by which the New International Economic Order can find fulfilment.

In this respect, it is obvious, and I am sure Mr Wilson will agree — that his proposals do not, and indeed do not intend to, constitute a comprehensive response to the Programme of Action for which the General Assembly has called. As the Secretary-General of UNCTAD has emphasised, that programme will remain frustrated unless approached through an integrated series of mechanisms. And, as has been demonstrated in a different context, it is not always that a step by step approach is either right or efficacious when seeking solutions to truly fundamental problems. We must, and I am sure I speak for all of us, thank Harold for the care with which he has advanced these proposals; but I owe it to him and to the meeting to indicate our initial anxieties about them."

Mr Burnham then went on to call for a much more general plan of action than that implicit in the Wilson recommendations and laid emphasis on the point that the basis of future developments must be a working towards the new international economic order which would ultimately rectify the current imbalances in the redistribution of world wealth.

The expert group set up at the conference was under the chairmanship of Mr Alister McIntyre of Jamaica and reported back to a meeting of Commonwealth Finance Ministers in Guyana in August 1975. Its report, entitled "Towards a New International Economic Order" went much farther than Mr Wilson's speech and included a proposal for indexation of primary commodity prices which would ensure that they rise in line with inflation.

In summary, the discussions at Jamaica indicated that governments of less developed countries were likely to take a hard line and would not be deterred by what they considered to be attempts to "buy them off".

Evidence from other quarters supports the view that the preliminary intention of the industrialised countries will usually be to placate the less developed countries. Even the much vaunted Lome Agreement between the European Economic Community and some 46 associated less developed countries has recently come in for criticism and some would see it as a mechanism by which the European states would increase their sphere of influence in Africa on the principle that a "divide and conquer" approach may be useful in the coming decades in an era of confrontation.

Thus we have a situation in which decades of neglect of the needs of less developed countries by the industrialised countries has been replaced by a sudden but strictly limited concern with the continued availability of their products. The less developed countries, however, do not appear willing to tolerate this approach and more overt confrontation seems likely.

Forms of Confrontation

In this context we can examine prospects for the two other Brooks scenarios, that in which the less developed countries bring into being a new international economic order and that in which their attempt fails. Predictions are difficult as there is ample evidence to support each scenario.

Taking that in which the industrialised countries maintain their position, we can list a number of advantages which they possess. Firstly they remain dominant in the world in economic terms, even after the success of the OPEC oil price rises. They have considerable economic power and would, themselves be in a position to indulge in "processor power" and "manufacturer power", withholding much-needed exports of industrial goods from recalcitrant countries of the third world unwise enough to indulge in producer power. Secondly they have an overwhelming technological superiority, of crucial importance in the matter of substitution of raw materials, replacement by synthetics, and the development of novel extraction processes involving supplies available within their countries. This is clearly true for energy resources with the possible development of fusion power, but is equally important for large numbers of mineral resources. Dr Banks has pointed out some of the facts of this situation but the advantage of the industrialised nations also extends to agricultural commodities. Many of the agricultural commodities of high value from third world countries are already susceptible to substitution by synthetics, but additional commodities, extending even to coffee, tea and cocoa could also be susceptible.

There is another, even more important, sense in which agricultural commodities reflect an inherent advantage for the industrialised countries.

It seems likely that the next few decades will see major food shortages, especially of food grains, affecting large numbers of less developed countries in Asia, Africa and even Latin America. The developed industrialised nations of North America, Europe and Australasia are also major producers of food grain, and, what is more significant, are major exporters of these commodities. They thus have a very important producer power potential of their own. In other words, food could become a major weapon in the process of confrontation between rich and poor nations.

Finally, the developed industrialised nations retain a dominating military power which would be available in the event of a military confrontation in the future between the "rich north" and the "poor south". In all then, the industrialised nations have a complex of powers at their disposal with which to respond to any development of producer power on the part of less developed countries.

Nevertheless, a "victory" for the consumers would not necessarily imply a return to the era of abundant and cheap commodities. What is equally — perhaps more likely is that some commodity producers, especially those of plantation crops, may opt for a greater degree of self-reliance, with food crops replacing export cash crops. Thus provided with more of the basic needs of survival, and much less dependant on the international trading system for them, it would become more possible for such countries to control their own future development.

In addition, those less developed countries still engaged in primary production for export to industrialised countries would have their hand strengthened through fewer suppliers and reduced supplies.

However, one caveat should be added. Any such fundamental change in economic life style as implied in moves towards self-reliance would not find favour with the elite groups in producer countries who need plentiful foreign exchange to maintain that conspicuous consumption of western artifacts to which they have become accustomed.

Brook's other scenario was one of successful producer power stripping the industrialised countries of their privilege. At first sight this seems to have little in its favour, but some powerful arguments of support can be developed. One concerns the question of resource availability. It is often argued strongly that talk of depleted resources is misplaced in that adequate supplies of raw materials will usually be maintained; for minerals by a combination of the discovery of new reserves, new ways of improving extraction and possible substitution, and for biological resources by improvements in production and also the development of synthetics.

Such arguments ignore two factors. One is that such possibilities are more likely to encourage rather than discourage exercises in producer power. In other words, third world resource producers will be encouraged to employ producer power while they are able. The second factor is that producer power can be exercised long before a resource becomes depleted as has happened with oil and rock phosphates in recent years. Even at a conservative estimate, the member states of OPEC have oil reserves sufficient for 20 to 25 years yet this did not hinder them in exercising producer power.

Another factor lending support to this scenario is the manner in which differences among less developed countries tend to decline when potential for producer power emerges. It would have been difficult to envisage, in 1970, that such a disparate group of nations as the members of OPEC could, within 3 years, achieve such a demonstrable degree of unity on the matter of the price of oil. Nations as differing in their culture and politics as Iraq, Saudi Arabia, Iran, Nigeria, Venezuela and Libya all succeeded in acting in a remarkably unified manner.

It is now evident that co-operation between OPEC states and other third world countries is increasing rapidly. This is demonstrated by the results of the Dakar meeting and also by the manner in which most third world countries have refrained from criticising the OPEC group, even at such major gatherings as the 1974 World Food Congress. This drawing together is also demonstrated by the increase in development assistance from OPEC member states. Total commitments from OPEC countries rose from $3000 million in 1973 to $16000 million in 1974 while actual disbursements were of the order of $5000 million in 1974, compared with $11300 million for the much larger group of industrialised countries which are members of the Development Assistance Committee of the Organisation for Economic Co-operation and Development.

This is not the place to argue the merits of "aid", perhaps one of the most blatant misnomers of the twentieth century. The motives of OPEC member states may have much more to do with securing for themselves the leadership of an OPEC/third world power bloc than with the basic needs of development, but it remains true that close links are being forged. We should not ignore the strong possibility of exercises in producer power by weak third world producer countries being backed by the financial resources of OPEC member states. Even by mid-1975 there were suggestions that the major UNCTAD initiative for a long term stable commodities programme might only go ahead through the support of the oil producers.

In summary we are suggesting that current evidence indicates it to be unlikely that the developed industrialised countries will unexpectedly indulge in any degree of cooperation with third world countries. Some kind of confrontation is more likely. Each of Brook's confrontation scenarios has arguments in its favour and given the number of factors involved it is difficult to predict with any confidence how this confrontation might develop. Much will depend on how far the producers and consumers form themselves into opposing blocs or allow themselves to become fragmented.

The OPEC group and other third world countries probably have an ultimate continuity of interest; the latter, certainly, realising how little they can expect from the industrialised countries, have little option but to hope that the OPEC star will prove more beneficient. So far, at least, their solidarity with OPEC has produced tangible results.

OPEC, on the other hand, may well reckon that friendly relations with other commodity producers is a useful hedge against the future. The nascent producer power of the third world could eventually be turned against OPEC itself. Further, if the OPEC countries succeed in using their financial resources to industrialise themselves, then their own increasing demands on the world's commodity supplies will enhance the possibilities of producer power.

The consumer position is more difficult to evaluate. The original oil crisis resulted in a very considerable degree of disunity as various European nations scrambled to conclude purely bilateral deals with OPEC countries. The underlying variations in import dependency also tend towards policy differences. Thus the EEC, considerably more dependant on commodity imports than the USA, has followed a noticeably softer line towards producers. However, despite these differences, it may be regarded as likely that bloc tactics on the part of producers will evoke similar tactics by consumers.

Given the considerable strength of such blocs, confrontation could be continued indefinately. On the producer side, a generally weaker power base, but perhaps rather less to lose, whilst on the consumer side an overwhelming superiority in military, political and economic terms partially balanced by the vulnerability created by complexity, affluence and a lesser sense of desperation.

In such a confrontation, we would suggest a fourth scenario as the one most likely, with *all* participants suffering. Ideally one might hope that the developed industrialised nations would then rapidly set about making the necessary concessions before the global situation became intolerable. Unhappily, the governments concerned are unlikely to be able to find the political means to act in such a manner. Pursuit of short term political advantage has tended to make it very difficult to engender in an electorate an appreciation of long term realities. Added to this, the inculcation of the notion that a high and rising standard of living is a congenital right will not readily be squared with the necessity for "levelling down" on a world scale. Finally, the impression created that the rich nations are really very generous to the poor, giving large sums in aid, will create by consequence a quite wholly chauvinistic outrage if the recipient poor appear to be biting the hand which so "generously" feeds them.

To add to all this, there is a further significant factor to be considered, that of unconventional warfare. If the processes of producer power do succeed in effecting a partial redistribution of world wealth towards the less developed countries, it is certainly probable that much of that wealth will be held by the relatively small rich elites in most third world countries. This would increase the disparities in wealth within these countries and we should expect an eventual increase in unconventional warfare. Such warfare whether practiced in Ulster, the Middle East or elsewhere, has become a feature of human activity in the late twentieth century, but has usually been allied to nationalistic motives. In any extension to a more general "army of the poor" it would be directed against these elites as well as the industrialised nations as a whole.

It cannot be stressed too strongly that, while countermeasures continue to be refined, industrialised societies are increasingly vulnerable to unconventional warfare. The use of the hi-jack, the proxy bomb and coercion in various forms have not yet been extended to even a limited nuclear capability, but such a development remains possible.

In conclusion, then, we suggest that if sane policies of cooperation are eventually effected, this will only happen after idiot policies have been *experienced* as paying no dividends. When the shocks and dislocations of an economic war have pruned the affluence of the rich nations and when the increasingly destructive activities of unconventional warfare have created a generally fearful atmosphere, then the benefits of a more orderly mode of conducting human relationships may become apparent.

Notes and References

[1] For a detailed account of the origins and early years of UNCTAD see "The Less Developed Countries in World Trade" by Michael Cutajar and Alison Franks. Overseas Development Institute, London, 1967.

[2] Guardian 23 May 1972.

[3] UN General Assembly resolution 2398 (XXIII) of 3 December 1968.

[4] "Limits to Growth" by D H Meadows, D L Meadows, J Randers and W Behrens. Earth Island, London, 1972.

[5] In "Human Ecology and World Development", edited by Anthony Vann and Paul Rogers. Plenum Press, 1974.

[6] Jugoslav State News Agency press release.

[7] Source: Financial Times, March, 1974.

[8] In "Human Ecology and World Development", edited by Anthony Vann and Paul Rogers, Plenum Press, 1974.

CONCLUSIONS

A feature of human activity frequently stressed by environmentalists is the "compression" of the time scale of such activity in recent history. Man evolved from a tool-making hominid to his present state in approximately 3 million years. The development of agriculture and consequent urbanisation commenced a mere 12000 years ago, the industrial revolution is barely 250 years old and we are now in an era of rapid population growth and increasing pressures on renewable and non-renewable resources.

Such an introduction is normally a prelude to an apocalyptic vision of man's downfall but our purpose here is merely to emphasise the recent acceleration in man's ability to affect the total environment in relation to the much slower rate of evolution of that environment. While this ability is not destabilising *per se,* there has long been a potential for such a phenomenon because man's ability to institute effective ecologically sensitive decision-making processes has been greatly outstripped by his technologically advanced culture. Indeed, the fragmentation of decision-making, its slowness and lack of long-term direction would seem often to preclude both the rational development and protection of resources and the planned development of the underprivileged areas of the world.

An essential aim of this book has been to suggest that we are now beginning to understand the manner in which man's increasing effects on the total human ecosystem are likely to affect human society over the next twenty-five years. As the initial contribution to this volume infers, if action over resource use is taken in the near future, a rational pattern of development may be possible in the long term. Suitable safeguards are, however, necessary with respect to two main factors in this process:

(a)   the adoption of technology of minimal environmental impact in terms of direct hazards, and

(b)   the protection of the global ecosystem, many factors of which require extensive study, to maximise stability.

Our contention is, however, that the necessary action seems unlikely to be forthcoming and that the most important trend in man's environmental affairs over the next twenty-five years will be conflict over the use of renewable and non-renewable resources.

We should not ignore the fact that such conflict is an old-established facet of human activity, but the main difference between past and future confrontation lies in the fact that our knowledge of global resources is increasing in accuracy. The size of the "cake" from which the "pieces" are to be drawn is becoming more clearly understood so that any forth-coming conflict is more likely to be definitive.

The less developed countries of the world will seek increasingly to redress the current imbalances in the global distribution of wealth through the medium of producer power. By demanding higher prices for the diverse raw materials they supply to developed industrialised countries they will attempt a redistribution of wealth sufficient to over-come the many problems of poverty from which they suffer. Ideally, if such a process were to work it would be accompanied by two other developments. Wealth would become more equitably distributed within the less developed countries and in the industrialised countries there would be moves towards a more conservationist life style, with heavy emphasis on the maximum recycling of resources.

Indeed the process by which producers attempt to influence market prices may do much to alleviate the fears of some ecologists who believe that certain key resources may become limiting in the near future. Price regulation by the volume control of raw materials is likely to extend the production cycle of resources so controlled. To some extent though, such savings would be offset by the increasing use of the same resources by the increasingly wealthy primary producers. Even so, the hoped-for result of producer power would be an era of international cooperation devoted to the conservation of the earth's natural resources and the increased well-being of its human population.

This is an ideal, and it presupposes a radical change in our present socio-economic systems in both developed and less developed countries. At present, the prospects are that moves towards producer power will produce confrontation at the international level and, even if successful, there is little to indicate that any benefits will accrue to those most in need. In this context, even the most ardent advocate of third world development must question producer power as a mechanism for develop-ment, adding as it does yet another series of international tensions to an already grossly imbalanced world system.

It is our contention that conflict over resources is the most likely development over the next twenty-five years but it certainly remains possible that such conflict may be avoided. Opinions vary concerning the outcome of the recent Seventh Special Session of the UN General Assembly, in that the results, on paper, indicated strong moves towards increased cooperation between developed and less developed countries. Ken Laidlaw, in a postcript to this volume, suggests that any optimism may be premature and we would emphasise that it may take several years to discern whether patterns of cooperation or confrontation are likely to evolve.

In endeavouring to understand events over the next few years, several important points should be considered.

1. OPEC was a special case; the world's most important commodity, subject to increasing demand and with exporting countries united in pursuit of a common goal and supported by considerable financial resources. All these factors gave OPEC real producer power. No other resource originating in less developed countries has this kind of latent power and very few third world resource producers will ever be in a position to withstand pressures from consumer interests unless they have external backing. This suggests that any link between OPEC and other third world resource producers is likely to be of the utmost significance.

2. The situation for a particular commodity can change in months or even weeks. All the reasons why producer power will not work for a particular raw material can count for very little if some unpredictable factor determines that demand suddenly outstrips supply. This, of course, works in reverse, and a sudden and unexpected decline in demand for a resource can quickly kill any potential for producer power for that resource.

3. It is all too easy to forget the many heavily populated countries which are unlikely to benefit significantly from producer power unless there is some unexpected and highly welcome sharing of any fruits of producer power among third world countries. Instability from this standpoint could be almost as disruptive as confrontation between producer and consumer countries.

4. Energy is fast becoming the international form of currency. While other commodity cartels may have some effectiveness, it will be the energy cartels (both present and future) which will ultimately dominate economic activity. The only escape from this lies within the field of self-sufficiency, with the use of renewable energy sources in particular. Most of these operate most efficiently on a highly localised scale and would seem to be one of the most fundamental keys to development in the third world, as well as an essential requirement for the developed industrialised world.

5. Finally, it is essential to emphasise the risk of increased unconventional warfare. If poverty continues on its present scale or if producer power meets with some success internationally but fails to heal inequalities within nations, then reaction from "the global ghettoes of the underprivileged" becomes not just possible but well nigh certain.

In conclusion there would appear to be still time to effect the changes necessary to prevent international conflict over resources, but the pace of events has moved so fast in the past three years that action is going to be required without delay. At the same time, disquiet over the ecological consequences of further development leads us to believe that the time is ripe for the reorganisation of human activity towards becoming sustainable in the long term. The mechanisms, in decision-making terms, by which such changes can be brought about are unclear, but as time goes on the choices open to man on a global scale decrease at an ever increasing rate.

"Stand still, you ever-moving spheres of heaven,
That time may cease and midnight never come."

Christopher Marlowe, 1564—1593

Part II

TRANSITION

INTRODUCTION

Early in 1974, when the long term implications of the activities of OPEC were beginning to be appreciated, a commentator in Britain suggested that we were living in "a world turned upside down". This is certainly a view expressed several times in the preceeding papers — that the latent power of the resources controlled by less developed countries could be sufficient to lead to a fundamental alteration in their relationships with the developed industrialised nations.

The aim of this book has been to examine some of the longer term trends inherent in the phenomenon of producer power, but it is essential also to look at some of the short term changes in attitude which have been in evidence in recent years. In order to assist in this process, we have collected together a number of speeches and documents from the period September 1973 to May 1975.

Starting with the Economic Declaration of the Algiers summit meeting of non-aligned countries in 1973, it shows the way in which a realisation of producer power developed from the manifest success of the OPEC model through to a more general call for a new international economic order early in 1974.

By 1975, third world leaders such as Algeria's President Houari Boumedienne has laid down a fairly clear plan of action, and while, at least in the area of commodities, UNCTAD continued to play a major role, one of the first major discussions on producer power and the new international economic order took place in Jamaica in May 1975 at the meeting of Commonwealth heads of government.

Relevant papers from all these sources are included here and help to show the change in thinking in international development studies which has taken place in recent years. It will probably be some time before the full significance of the emergence of producer power in 1973—5 can be assessed and our purpose in publishing these documents is to help to further that process.

THE ECONOMIC DECLARATION OF THE ALGIERS SUMMIT MEETING

SEPTEMBER 1973

Introduction

Since 1960, four meetings of heads of state of non-aligned countries have taken place. The most recent meeting was in Algiers on 5 to 9 September, 1973, a few months after UNCTAD III and the UN Human Environment Conference in Stockholm and at a time when it was becoming clear that the oil producing countries were about to engage in a process of bargaining over the price of their principle export.

The Economic Declaration from the summit meeting was, in many respects, typical of the kind of declaration made at such meetings, calling for a new sense of urgency in matters of world development, with the major responsibility resting with the developed industrialised nations. Its significance, however, lies in the attention paid to the possibilities for producer power, one of the first occasions on which such consideration had been given at a major meeting of third world heads of state.

FOURTH CONFERENCE OF HEADS OF STATE OR GOVERNMENT
OF NON-ALIGNED COUNTRIES

ALGIERS, 5–9 SEPTEMBER 1973

ECONOMIC DECLARATION

Introduction

The Heads of State or Government of Non-Aligned Countries made a
detailed review of the development of economic and social conditions in
developing countries in the world context prevailing since the Lusaka
Conference on Trade and Development, the United Nations Conference
on the Human Environment, the preparation for the multilateral trade
negotiations, the reform of the monetary system and the important
on the Human Environment, the preparation for the multilateral trade
negotiations, the reform of the monetary system and the important
Conference of Foreign Ministers of Non-Aligned Countries held at
Georgetown.

They noted that the trend in the international situation towards
détente, for which the non-aligned countries have always striven and
which is a positive factor for the strengthening of peace in some parts
of the world, has had little appreciable effect on the development of the
developing countries and on international co-operation. The Heads of
State or Government therefore consider that the increasing trend towards closer economic relations between developed countries should in
no way adversely affect the basic interests of developing countries.

I   The struggle against imperialism

The Heads of State or Government of Non-Aligned Countries
noted that imperialism is still the greatest obstacle to the emancipation and progress of the developing countries which are struggling to achieve levels of living compatible with the most basic
standards of well-being and human dignity. Imperialism not only
hampers the economic and social progress of developing countries,

but also adopts an aggressive attitude towards those who oppose its plans, trying to impose upon them political, social and economic structures which encourage alien domination, dependence and neo-colonialism.

This situation derives from a systematic policy pursued by imperialism everywhere which remains unchanged even though its implementation may assume different forms depending on the circumstances of the time and the place. In particular, it must be pointed out that, in addition to being an infringement of the principles of sovereignty and independence, imperialist policy sometimes has the character of open aggression against the economies of peoples which refuse to submit to alien domination. This policy even resorts to the use of force and the unleashing of criminal wars such as those which are still affecting the peoples of Indochina and the Arab peoples of the Middle East.

Many countries are still subject to imperialist domination and neo-colonialist exploitation, which constitute a threat to the sovereignty of States and hamper the development of the peoples concerned.

Moreover, some peoples are still victims of direct colonization and apartheid which deprive them of their fundamental rights to sovereignty and independence and prevent any possibility of development.

This situation accounts for the considerable disparities which exist between the industrialized countries and the underdeveloped world.

Thus the developing countries, in general, are still subject directly or indirectly to imperialist exploitation. Colonialism and imperialism have been unable to withstand the vast political liberation movement marked by the historical turning point of Bandung, but they have adapted themselves in order to perpetuate in another form their stranglehold on the resources of the developing countries and to ensure for themselves all kinds of privileges and guaranteed markets for their manufactured products and services.

Policies have been implemented which are based on the use of overt and covert economic aggression, as is illustrated by the manifold and increasingly pervasive activities of transnational and monopolistic commercial, financial and industrial companies.

In their struggle to achieve independence, economic development and full equality in international relations, the non-aligned countries, individually and collectively, with the support of all the progressive forces in the world, are effectively resisting imperialist aggression, and are thus emerging as a major force in the struggle against imperialism throughout the world.

II   Economic Situation of the Developing World

In view of this increasingly alarming situation, many compelling factors, the most significant of which is the determination of the peoples to free themselves from any form of alien domination by taking their destiny into their own hands, have led the international community to elaborate various policies aimed at establishing a new type of international economic relations.

The determinations of the vast majority of the developed countries to perpetuate the existing economic order for their sole benefit, without regard for the wishes of the developing countries, has virtually thwarted all attempts at progress. The failures of the First Development Decade and the unsatisfactory implementation of the recommendations of the third session of the United Nations Conference on Trade and Development, together with the disappointing results of the first three years of the current Decade, have already jeopardized the achievement of the objectives of the International Development Strategy.

III   Evaluation of the International Development Strategy

The developing world, which accounts for 70 per cent of mankind, subsist on only 30 per cent of world income.

Of the 2,600 million inhabitants of the developing world, 800 million are illiterate, almost 1,000 million are suffering from malnutrition or hunger, and 900 million have a daily income of less than 30 U.S. cents.

In the light of all these considerations, estimates up to 1980 are extremely pessimistic. Assuming that the targets set for the Second Development Decade can be achieved, and this is by no means certain, gross national income per capita in the developing countries would increase by only 85 U.S. dollars as against 1,200 U.S. dollars in the industrialized States. By the end of the present decade, average annual *per capita* income will be 3,600 U.S. dollars in the developed countries, but only 265 U.S. dollars in the developing countries.

This failure of the International Development Strategy is due both to the lack of political will on the part of the developed countries to take urgent action, and to the inadequacy of the *growth* target in relation to the real needs of the developing countries.

Indeed, the necessary international co-operation has been lacking. The attitudes of the Governments of some developed countries, and the behaviour of transnational firms and other monopolies benefiting from the plundering of developing countries, have not contributed to the creation of an external economic situation in line with the objectives of the International Development Strategy.

Other factors are the inflationary rise in the cost of imports, the pressures on the balance of payments caused by transfers by private foreign investors, loan repayments and the heavy cost of external debt servicing and the aggravating effects of the international monetary crisis.

The arms race and the competition for the conquest of space continue to absorb large sums of money, whereas assistance through international multilateral co-operation is becoming smaller and smaller in relation to the growing needs of developing countries.

The numerous projects intended to enable the developing countries to benefit in an organized manner from the results of scientific research and technological progress have not even begun to be seriously implemented, while the developed Western countries continue to monopolize the services of large numbers of highly qualified personnel, especially scientists and technicians, from the developing countries.

Clearly, however, only a proper conception of development which is based on the requisite internal structural changes particular to each country, and which encompasses growth in all the key sectors, will enable our countries to achieve their development targets. This process is inseparable from another process, social in nature, which calls for the highest possible employment levels, income redistribution and the over-all solution of problems such as health, nutrition, housing and education. It is equally obvious that these aims can be achieved only with the conscious and democratic participation of the masses, which are the determining factors in any national endeavour to achieve dynamic, effective and independent development.

IV   Trade and Monetary Problems

The Heads of State or Government noted that the already modest share of the developing countries in world trade is continually decreasing, while the terms of trade are constantly deteriorating.

The share of the developing countries in world trade declined from 21,3 per cent in 1960 to 17,6 per cent in 1970.

The Generalized System of Preferences automatically excludes the main agricultural products and imposes strict controls on the import of all products considered as sensitive by the developing countries; in addition, it is not applied by all countries.

The recent increase in the price of certain raw materials has not benefited the developing countries as a group, since import prices have increased even more, and the profits resulting from the rise in the price of raw materials have been made by the transnational companies. The trade situation of the developing countries has been aggravated by the international monetary crisis, of which they are bearing the full brunt, although they are in no way responsible for it.

The transfer of resources from the developed to the developing countries has continued to decrease, while the volume of the latter's external debt has quadrupled during the last decade and is now over 80,000 million U.S. dollars.

Furthermore, there has been no improvement in the terms of development finance.

The Heads of State or Government noted that economic power has hitherto been used in trade negotiations to frustrate the aspirations of the developing countries. They therefore regard the forthcoming multilateral trade negotiations as of great importance for reversing the adverse trends in the trade of developing countries. They agreed that the non-aligned countries and other developing countries should take a united stand in the negotiations and aim at universal acceptance of the principles of equity in international relations. They strongly believe that the multilateral trade negotiations should pave the way for a new and more equitable international division of labour. To this end, the negotiations should aim at:

—Ensuring for developing countries net additional benefits, an increased share in world trade and diversification of their exports;

—Ensuring that other bodies place emphasis on complementary objectives and measures to enable developing countries to obtain maximum benefits from the negotiations;

—Winning acceptance of the principles of non-reciprocity, non-discrimination and preferential treatment in respect of developing countries;

—Extending the Generalized System of Preferences;

—Ensuring that any erosion resulting from the negotiations is compensated;

—Ensuring that the co-ordinated approach to trade and monetary problems takes the fullest account of the special interest of developing countries;

—Ensuring that preferential treatment in respect of developing countries is included in any reform of international exchanges and of the rules of GATT.

The reform of the international monetary system concerns the developing countries in all its aspects and to the highest degree.

Because of the basic principles governing it and the manner in which it operates, the monetary and financial system devised at BRETTON WOODS has served only the interests of some developed countries. The efforts made by the developing countries to bring about a progressive adjustment of the BRETTON WOODS system, in order to take account of their specific needs, have not been successful. This clearly illustrates the lack of political will on the part of certain industrialized countries to establish and promote genuine co-operation between developed and developing countries within the framework of the international financial and monetary system.

The new international monetary system, in the establishment and operation of which the developing countries should participate on an equal footing, should be universal, should guarantee the stability of monetary flows and the conditions of financing international trade, and should take into account the specific situation and needs of developing countries on the basis of preferential treatment.

V   Special Measures in Favour of the Least Developed Countries, Including Land-locked Countries

The Heads of State or Government, viewing with concern the persistent stagnation in the economies of the least developed countries, consider that the international community should intensify the special support given to these countries in the agencies of the United Nations system by increasing the volume of aid and providing them with financial and technical assistance in all fields of development, including communications and diversification of exports.

They further consider that special international assistance should also be provided to the land-locked developing countries to enable them to overcome their geo-structural handicap and to derive full benefit from the resolutions adopted by United Nations agencies in this regard.

In this connexion, the problems of land-locked countries surrounded by countries against which economic sanctions have been applied by the United Nations deserve special attention.

VI   Food Problems

In view of the catastrophic scale of the food crisis in vast areas of the world, especially in the Sudano-Sahelian region of Africa, which is aggravating the food shortage that has continued for fifteen years without improvement, it is imperative that the international community adopt as a matter of extreme urgency the measures dictated by this situation, which is now coupled with the unchecked rise in the price of staple products.

The Heads of State or Government consider it necessary:

—to adopt more effective solutions than those at present proposed in the field of international co-operation concerning staple products;

—to rescind the restrictive measures relating to production and stocks, which have a highly detrimental effect on the volume, and substantially increase the price, of certain agricultural products of the developed countries which are essential to the developing countries.

VII   Sovereignty and Natural Resources

In view of the seriousness of the problems confronting them, the developing countries realize more than ever before the vital necessity of making every possible effort to consolidate their national independence and reinforce their struggle by challenging imperialist and neo-colonial exploitation structures, and by organizing co-operation and solidarity with one another in intercontinental and regional organizations. The action taken in the non-aligned countries after the Belgrade, Cairo, Lusaka and Georgetown Conferences, the decline of colonial and neo-colonial groupings, the strengthening of the unity of action of the Group of 77, particularly on the basis of the Charter of Algiers and the Lima Declaration, and regional co-operation and integration activities, are all steps marking the transition from the passive submission of claims to the affirmation of the developing countries' determination to rely first and foremost on their own resources, individually and collectively, to take over the defence of their fundamental interest and to organize their development by and for themselves.

The Heads of State or Government, while recalling the inviolable principle that every country has the right to adopt the economic and social system which it deems most favourable for its own development, reaffirm the inalienable right of countries to exercise their national sovereignty over their natural resources and all domestic economic activities.

Any infringement of the right of effective control by any State over its natural resources and their exploitation by means suited to its own situation, and having regard to ecological good neighbourliness, including the right of nationalization and the transfer of property to its nationals, is contrary to the Purposes and Principles of the United Nations Charter and hampers the development of international co-operation and the maintenance of international peace and security.

The Conference gives its unreserved support to the application of the principle that nationalization carried out by States as an expression of their sovereignty, in order to safeguard their natural resources, implies that each State is entitled to determine the amount of possible compensation and the mode of payment, and that any disputes which might arise should be settled in accordance with the national legislation of each State.

The non-aligned countries give their ready and unreserved support to the developing countries and to the territories under colonial domination which are subject to boycott, economic aggression or political pressure and are struggling to recover effective control over the natural resources and over economic activities which are still under foreign domination.

In this connexion, the Heads of State or Government recommend the establishment of effective solidarity organizations for the defence of the interests of raw material producing countries such as the Organization of Petroleum Exporting Countries (OPEC) and the Inter-Governmental Committee of Copper-Exporting Countries (ICCEC), which are capable of undertaking wideranging activities in order to recover natural resources and ensure increasingly substantial export earnings and income in real terms, and to use these resources for development purposes and to raise the living standards of their peoples.

In this connexion, the results obtained in the hydrocarbons sector, which was previously exploited for the sole benefit of the transnational oil companies, demonstrate the power and effectiveness of organized and concerted action by producing and exporting countries.

Similarly, the determination of an increasing number of developing countries to terminate treaties, agreements and conventions imposed on them by force and violence, is producing increasingly positive results. This process should be extended, accelerated and co-ordinated in Latin America, Asia, Africa, the Middle East, and in developing countries elsewhere, in order to strengthen solidarity among the developing countries, reverse the trend towards a deterioration of their situation and secure the establishment of a new international economic order which would meet the requirements of genuine democracy.

The non-aligned countries decide to use all possible means to ensure that the global approach for the achievement of the aforementioned objectives is accepted by the international community, which will have to take the fullest possible account of the provisions contained *inter alia* in the Charter of Algiers, the Lusaka Declaration, the Lima Declaration and the Georgetown Action Programme.

VIII   Transnational Companies

The Heads of State or Government denounce before world public opinion the unacceptable practices of transnational companies which infringe the sovereignty of developing countries and violate the principles of non-interference and the right of peoples to self-determination, which are basic prerequisites for their political, economic and social progress.

The Conference also recommends that arrangements be made for joint action by the non-aligned countries in regard to transnational companies, within the framework of a global strategy designed to transform qualitatively and quantitatively the system of economic and financial relations which subordinates the developing countries to the industrialized countries.

IX   Transfer of Technology

The Heads of State or Government of Non-Aligned Countries recognize the need for developing countries to bridge the gap which separates them from the industrialized world in the field of technology.

They are therefore aware of the need to create or improve a technology adapted to the needs and realities of their countries by intensifying their own research efforts and by profiting from the mutual experience of the non-aligned countries. In addition, they agree to continue the struggle within international organizations to obtain easier and less costly access to modern technology and for the adoption of an international code of conduct governing transfers of technology from the developed countries to the developing countries, taking due account of the latter's independence.

X   Co-operation among Developing Countries

The Heads of State or Government are convinced that there are great opportunities for economic, trade, financial and technological co-operation among developing countries, particularly in the political and economic conditions of the present day world.

They accordingly recommend that everything possible should be done, through individual or collective action, with a view to strengthening this co-operation.

XI   Co-operation between Developed and Developing Countries

The trend in the international situation towards *détente* between socialist and capitalist countries has had only limited effects on international co-operation in favour of development. On the contrary, there is a marked trend towards increasing the complementarity of the economies of developed countries, and consolidating their economic groupings, which are co-operating more and more closely with one another, often neglecting the major interests of developing countries.

The Heads of State or Government, while noting the negative nature of the policy of the developed market-economy countries, nevertheless observed with hope the favourable approaches that some of those countries are currently adopting with regard to development problems. They have also noted the increase in contacts and the closer co-operation between developed countries with different economic and social systems. This trend should be extended to the entire international community so as to provide a basis for the evolution of a new pattern of economic, trade and technological relations based on national complementarities and creating conditions for a more complete development in accordance with their needs and potentials. The Heads of State or Government strongly urge all States Members of the United Nations to keep in view these broad objectives when developing their mutual relations.

XII   Environment

The Heads of State or Government reaffirm their concern to ensure that the extra cost of environmental programmes should not provent the fulfilment of basic development requirements, and they regard economic backwardness as the worst form of pollution.

They recognize that developing countries have their own environment problems which differ from those of developed countries and which require the attention of the international community.

While the developed countries' environment problems may be partly solved by deconcentrating pollutant production units, those of the developing countries require the provision of additional resources by the international community.

The Heads of State or Government also reaffirm that all environmental co-operation between developed and developing countries should be additional to what is already being provided in the form of development aid.

They stress that environmental measures taken by one State should not adversely affect the environment of other States, or zones outside its jurisdiction.

The non-aligned countries consider it necessary to ensure

effective co-operation between countries through the establishment of adequate international standards for the conservation
and harmonious exploitation of natural resources common to two
or more States within the framework of the normal relations
existing between them.

They also believe that co-operation between countries interested in the exportation of such resources should be developed
on the basis of a system of information and prior consultations
within the framework of the normal relations existing between
them.

Co-operation between developed and developing countries in
the environmental field requires that the former de-mine the
territories which they had mined during previous wars and acts
of aggression, since these mines are a source of pollution in a
number of developing countries.

XIII   Charter of the Economic Rights and Duties of States

The non-aligned countries believe that the United Nations
General Assembly should, at its twenty-eight session, give priority to the elaboration of the Charter of the economic rights and
duties of States.

This document should reflect the economic aspirations of
states which are struggling to achieve over-all development, as
well as those of the international community as a whole.

XIV   Preservation and Development of National Cultures

It is an established fact that the activities of imperialism are not
confined solely to the political and economic fields but also
cover the cultural and social fields, thus imposing an alien ideological domination over the peoples of the developing world.

The Heads of State or Government of Non-Aligned Countries
accordingly stress the need to reaffirm national cultural identity
and eliminate the harmful consequences of the colonial era, so
that their national culture and traditions will be preserved.

They consider that the cultural alienation and imported civil-
ization imposed by imperialism and colonialism should be countered by a repersonalization and by constant and determined
recourse to the people's own social and cultural values which
define it as a sovereign people, master of its own resources, so
that every people can exercise effective control over all its national wealth and strive for its economic development under conditions ensuring respect for its sovereignty and authenticity, and
peace and genuine international co-operation.

THE SIXTH SPECIAL SESSION OF THE UN GENERAL ASSEMBLY

Introduction

The sixth special session of the UN General Assembly was the first such session to be exclusively concerned with international economic affairs. It was summoned following an initiative of President Boumedienne of Algeria and took place in New York in April 1974 with the purpose of discussing problems of raw materials and development.

We reprint here an account of the special session by Barbara Ward originally published in "The Economist" on 18 May 1974, together with the two major documents arising from the session, the Declaration on the Establishment of a New International Economic Order and the Programme of Action which accompanied that declaration.

As with the economic declaration from the Algiers summit, so we find that the two documents from the sixth special session are largely similar in content to other previous declarations on international development. The two essential differences are, firstly, that more attention was paid to the role of resources in achieving this new economic order, and, secondly, that the documents arose from a major meeting of the international community of nations where the previously weak less developed countries were beginning to appreciate their potential strength.

FIRST, SECOND, THIRD AND FOURTH WORLDS

Barbara Ward

When the special session of the United Nations General Assembly, summoned on the initiative of President Boumedienne of Algeria to discuss "the problems of raw materials and development", closed on May 2nd, it left behind the feeling that possibly something new had taken place. Mr Henry Kissinger called the session part of an "unprecedented agenda of global consultations in 1974" which implied "a collective decision to elevate our concern for man's elementary well-being to the highest level". Britain's chief representative at the United Nations, Mr Ivor Richard, put it rather more simply. He said: "Things will never be the same again".

Yet the change could well have escaped the casual observer. If we count the marathon discussions held at the three Unctad conferences and add in the 1970 debates in the UN assembly on a "strategy for the second decade of development", it cannot be said that most of the points at issue during the special session had not been discussed before. At least half of President Boumedienne's speech at the opening of the session covered familiar ground: the bias against the poor nations in the world economy, their unfavourable terms of trade in the 1950s and 1960s, the pile-up of their debts—now standing at $80,000m—the involuntary depreciation of their reserves, their sense of powerlessness at the highest level of international decision-making in investment and monetary affairs.

Nor did the two working documents prepared for the special session by the developing countries' Group of 77—a declaration of principles and a programme of action—prove to be strikingly different from earlier pleas and proposals for international action. The principles denounced exploitation, foreign occupation, colonialism and apartheid, declared the nations' right to control their own resources and to nationalise them if necessary, and asked for control over multinational corporations, for recognition of "producers' associations", for universal representation in international bodies and a speedier implementation of the policies agreed on for the second development decade.

These policies, set out in the programme of action, cover the customary rubrics of more aid, more investment in developing nations' industry, more preferential treatment in trade, higher raw material prices and monetary arrangements geared to the needs of the poor. If listeners to the lengthy speeches and readers of the flow of documents acquired a slight sensation of having heard much of the debate before, a rapid perusal of earlier documents would show their suspicions to be justified. "What I tell you three times is true" is, of course, very largely the case in relation to the developing world's problems. But it does not always make for easy listening.

Moreover, the end of the session had a melancholy resemblance to the close of earlier consultations. The principles and the plan of action, somewhat modified in three weeks of debate, were carried by a fairly exhausted assembly with enough reservations by the rich nations to make the value of the whole exercise look fairly dubious. And a last-minute resolution on emergency aid submitted by the one nation whose adherence to some plan of action is indispensable—the United States—was pushed aside on the grounds that the debate was already over. Thus the scenario seemed sadly familiar—a paper victory for a co-operative world economic strategy with neither resources, instruments nor political will to carry it through.

Yet it was at the close of the session that Mr Ivor Richard made his remark. If, indeed, "things will never be the same again", something reasonably decisive must have been going on under the interminable exchange of set speeches and familiar grievances. And if one looks a little below the surface, what appears is not the old ritual performances but a series of interlocking changes which affect virtually every aspect of the international economy.

In a Class by Itself

Perhaps the most fundamental change is the degree to which the session suggested a loosening and changing of all the supposed "blocks" or "worlds" into which the international economy has been divided and upon which so much of past rhetoric has been based. The conventional image of recent years has been of a first world of developed market economies, a second world of "socialist" states, and the "third world" of the developing nations. Not one of these distinctions looked really sustainable as the debates went on. Take the developed market economies. The United States, the European Economic Community and Japan said some of the same things—that aid should be maintained and special efforts made for the most hard-hit states, that "orderly and co-operative" arrangements be considered to ensure stable raw material prices, and new efforts be undertaken to increase scarce supplies, particularly of food and fertilisers. But it does not take any very close examination of the various memoranda on prices, trade and the balance of payments prepared by the UN system for the special session to realise that the United States is, now more than ever, entirely in a class by itself.

It is far less dependent upon oil imports than other market economies. As a proportion of its use of energy, oil imports represent only 13.5 per cent of the total, imports from North Africa and the Middle East only 2 per cent. For western Europe the figures are 59 per cent and 47.4 per cent respectively, for Japan 72.6 per cent and 57.4 per cent. Even more striking are the vast gains the United States has made in the last two years in the world trade in grain. On April 25th, in the middle of the special session, the American Secretary of Agriculture, Mr Earl Butz, pointed out in Washington that American petroleum imports in 1973-74 were comfortably covered by an increase of $9 billion in food exports. Of this, $7 billion represents the tripling of the price of grain and—as many developing nations were quick to notice—$2 billion have been earned by food sales to the poorer lands.

Mr Butz also warned the world that there would be "no more storage at the expense of the American taxpayer" and that those who wanted grain had better get into line to buy it and store it themselves. This clearly implies little or no future concessionary food sales and a determination on the part of some sections of the American Administration to see that at least one raw material, food, is not subject to international agreements or constraints.

The very different degrees of bargaining power enjoyed by the United States on the one hand and, on the other, the resource-hungry nations of the EEC and Japan, did not surface at the session. Nor was there any repetition of the open split between France, seeking to represent western Europe's different dilemmas over oil, and the United States trying to concert a move by the whole developed world to roll oil prices back. But neither was there any observable first world strategy designed to deal with the critical issues. As for its being a much feared and much denounced "cartel" of the rich, the group did not seem to function at all.

Great Disorder Under Heaven

Nor, it must be admitted, did the so-called second world. In fact, only here did the rhetoric of denouncing colonialism and imperialist domination really have the ferocious bite of passionate conviction:

> Under the name of so-called "economic cooperation" and "international diversion of labour", it uses high-handed measures to extort super profits in its

> "family" . . . Its usual practice is to tag a high price on outmoded equipment and sub-standard weapons and exchange them for strategic raw materials and farm produce of developing countries. Selling arms and ammunition in a big way, it has become an international merchant of death.

It might be illuminating to offer odds on which power is doing the denouncing and which is the government denounced. The denouncer is, of course, the People's Republic of China, with the Soviet Union on the whipping block. Possibly as a result of this internal contradiction, the Russians preserved a profile at the session low enough to recall the vaunted "low posture" of the Japanese. Meanwhile the Japanese concluded with them a large investment programme in Siberia in which Japanese investment and technology would open up, for Japanese use, Siberian minerals and timber resources.

As for the Chinese, they repeated their old Unctad stance. They applauded all efforts by developing peoples "to win or defend national independence, develop the national economy and oppose colonialism, imperialism and hegemonism". They backed the principles and the action programme. They initiated no proposals of their own, and allowed it to be understood that their labour-intensive, energy-conserving techniques represent a workable version of the future in developing lands, provided the peoples beyond the fringe are prepared to learn from the celestial model. Even the language of marxism seemed less potent for them than traditional Chinese expressions. They described the crisis as a time of "great disorder under heaven", and the passing of imperialism as an outcome dictated by the turning "wheel of world history"—an analysis which the facts do not necessarily contradict but which seems sufficiently far from the language of marxist orthodoxy.

If the first and second worlds did not function as groups, the third managed rather better. In momentum, language and intent its members more or less held together. One reason was undoubtedly the sense of vicarious strength many of them derived from the oil producers' ability to take an extra $66 billion from the industrialised states in a single year—a figure which may be compared with the $19.6 billion earned from all developed sources, public and private, by all the developing nations in 1972. Another reason was the degree to which the oil states, with Iran at their head, have increased their own offers of aid in the last year. The sum of firm and less firm offers is now in the neighbourhood of $4 billion to $5 billion. Still a third reason could lie in the belief of other raw material producers that they, too, may be able to profit by higher prices achieved through group action. Yet this hope also illustrates the underlying lack of cohesive interests in the so-called third world. Far from hoping to share in such a bonanza, the poorest and most populous states stand to lose by high prices in almost every way.

The Haves and Have-Nots

The raw materials lottery has, in fact, created at least four different types of developing state. Among the oil producers, the states of the Arabian peninsula—Saudi Arabia, Kuwait, Abu Dhabi and Qatar—with a total population of only 9m, have received additional revenue of the order of $33 billion since 1973 (Abu Dhabi's per capita income from oil alone is now $43,636). They have become rich absolutely, on a par with industrialised states and must, above all, worry about a rational use of so much wealth. The more populous oil producers, with over 270m inhabitants, have received added revenue of $31 billion. They could use every dollar on internal development. Another group of developing states—China, Colombia, Mexico, Bolivia—are more or less self-sufficient in oil.

Others, Morocco with its phosphates, Malaysia with its rubber, Zambia and Zaire with copper, are doing quite nicely on exports of commodities other than oil, some of which have doubled in price. A group of more developed states—from Brazil and Mexico to South Korea and Singapore —can hope to pass price increases for their manufactured exports on to their customers. Or they can secure access to the world's capital markets. But Mexico could lose heavily in the area of services—tourism or the export of workers—if recession sets in in the United States. So could the poorer Mediterranean countries if Europe stumbles.

Finally we come to the poorest developing states: the whole of the Indian subcontinent, tropical Africa, the Caribbean and parts of tropical Latin America. With this group, everything has gone wrong. They are importers of fuel, of food, of fertilisers. They have little access to capital markets. Some have no income either from tourism or from money sent home by migrant workers. And their products, notably tea and jute, have not gained in price. These are the states to feel the chief impact of the extra costs incurred by developing nations who do not export oil—a clear $15 billion for oil, $5½ billion for food and fertilisers, a doubling of prices for other materials and for many manufactured imports, and all this in a single year. So dire is the condition of the poorest countries, so distinct are they in deprivation from all the rest, that at this session the term "the fourth world" became common currency in describing their condition. This is the world's immediate disaster area, where famine is already present—in parts of west and east Africa—and could become inescapable in wider regions in 1975.

Clearly there are stark divisions of interest within the group of developing nations. The poorest countries, together with the less disadvantaged developing states who produce no oil, would seem to have the strongest inducement in securing lower petroleum prices. Theoretically, they could have been persuaded to join, say, in an American or joint first world initiative to put pressure on the oil producers. Equally they might have looked with dismay at a future in which producers of essential minerals set up price-boosting cartels from which they would be excluded simply by their lack of resources. Again, they might have broken ranks in order to seek particular agreements on aid and technical assistance with the developed states.

In the event, they held together and scored some pyrrhic victories in passing plans and principles at the end of the session. But the show of unity hardly papered over the total divergence of opportunity and policy between, say, a Saudi Arabia with 7.8m citizens and an oil revenue sextupled to $19.4 billion in less than two years and an India with 600 m people and total reserves before the oil crisis of only $1,355m.

Hanging Together

Yet it can be argued that the very fluidity of all the groups—the sense of changing interests, of uncharted possibilities, of new risks and new hopes—explains the fact that, after all, the session did mark a clear break with past UN performance. Like confused thoughts searching for an organising idea or floating molecules in wait for a catalyst, the underlying interests were too divergent to impose their own pattern.

And then a catalyst did appear—a growing realisation that, without emergency action, the poorest nations, the fourth world, would simply run out of reserves by midsummer and could with their bankruptcies set in motion the possible downward spiral of collapsing markets which, in 1929, finally engulfed the whole of the world economy. The prospect of hanging was present at Turtle Bay and wonderfully concentrated the the delegates' minds. From this focus of apprehension there grew a programme of action precise enough, urgent enough and sufficiently representative of all interests to justify the belief that "things would never be the same again".

At the centre of the programme is the concept of immediate emergency aid for those hardest hit by the jump in prices. The Shah of Iran had proposed a fund of $3 billion for the next year, the figure given by the World Bank as the minimum need for additional aid to the poorest states. Of this fund, $1 billion would come from the oil states and the rest from the industrialised nations. During the session itself, no firm pledges were given. Indeed, the inability of the grain-rich Americans to come up with any precise offer in the wake of Mr Kissinger's eloquent address explains why their last-minute proposal of a $4 billion fund for 18 months, with the United States doing its "fair share", was not even considered. But the principle of the fund was accepted and procedures established to make it a reality.

These procedures are as important as the agreement on the need for emergency funding. Hitherto the international economy has lacked any effective centre of impetus and strategic thinking. The World Bank, the

IMF and Gatt have been type-cast as the instruments of the rich market economies. Unctad, UNDP and some of the UN technical agencies are held to be more friendly to the poor. "Socialist" countries are represented here and there in the spectrum, but do not, in the main, patronise the "rich nations' organisations". The arrangements proposed after the special session could begin to change all this.

Once more, it is the model of the catalyst that comes to mind. The UN Secretary-General is empowered to make the appeal for emergency donations, to consult with a representative ad hoc group of governments in laying down conditions of eligibility, and to use the various agencies of the UN family—including the World Bank, the International Monetary Fund and the regional banks—to secure timely disbursement. He is asked to assess the quality and type of aid that is offered and monitor both its flow and the changing pattern of need. Mr Waldheim is also asked to take the initiative in seeking to establish a more permanent special fund at the beginning of 1975 to ensure that extra financial resources are available for the rather longer run—the World Bank reckons the need for 1975 at about $5 billion to $6 billion.

Reports on these activities are to go to the Economic and Social Council, the single universal instrument of the world community in the economic field. It is reported that the secretary-general has already invited Dr Raul Prebisch, the founder-father of Unctad, to lead the emergency operations—a move that will reassure the poor—and has secured the energetic adherence of the World Bank and the IMF—a condition that should mollify the rich.

The effectiveness of this operation depends, of course, on the speed with which the needed $3 billion can be secured. But the Iranians and the Algerians are giving a lead among the oil states; indeed, the patient chairmanship of Mr Hoveida of Iran during the session was a vital factor in allowing the catalysing impact of disaster to have its full effect. The EEC is expected to play its part and the hope is that, in spite of the clumsy handling of the American resolution, the American "fair share" will be spelt out and will prove to be $1 billion of aid in cash and in the food which the United States alone is in the position to supply. Mr Kissinger's interest in the UN conference on food next November—which he himself proposed—is felt to provide some assurance that the American offer will be validated.

Follow Up

This coming conference on food is only one among many. The population conference is due in August, next year's UN assembly will debate strategies for development, a conference on human settlements follows in 1976 and the French have proposed a UN conference on energy. The link between the new unit in the secretary-general's office and all these consultations is that it could develop into a more formal centre of stimulus and prediction able to serve an ongoing dialogue of the world with itself about its collective predicament.

Again and again during the debates of the special session, delegates from a wide variety of backgrounds and interests called for more information, more advice, more strategy, more wisdom—if that were possible—from the UN system itself. The suggestions included advisory councils, a group of "wise men"—the French and West German proposal—and a high level unit for assessing and monitoring aid and need. Other plans were rather more concrete. A number of delegations, including the Japanese, asked for a strengthening of the UNDP's fund for natural resources to enable it to plan, prospect and initiate action in the field of needed raw materials. Mr Kissinger made a widely-supported proposal for a world fertiliser institute to encourage the output of fertiliser, make it technically more efficient in developing lands—where plants running at a third of capacity are too often the rule—and to undertake research into new fertilisers and alternative methods of making existing ones. There was discussion, too, of an international effort in energy research to ensure supplies as the world's reserves of fossil fuels continue to dwindle.

And here perhaps we encounter the deepest reason for believing that the special session could mark a new beginning in international affairs. It was the first assembly to see surface, in unmistakable fashion, the chill possibility that the old idea of a "trickledown" of wealth from rich nations' constantly expanding resources on a scale sufficient to produce a succession of take-offs among the poor may not be a workable solution to the problems of development in the decades ahead. The expanding resources may simply not be there. So the issue is not simply the immediate one of rescuing the poorest nations from imminent bankruptcy. It could be the more alarming question of how developed and developing peoples are to survive in a planet where what Mr Kissinger calls the "elementary wellbeing" of all peoples, or Dr Walter Scheel "the humanisation of mankind", can be secured only by some restraint, some sacrifice of "gadgetry and over-consumption"—the phrase is President Boumedienne's—on the part of the already rich.

In the short run, this confrontation is a fact. A restored world food reserve—which the session virtually unanimously proposed—can be set up this year only if American food consumption is somewhat reduced. (Since medical authorities in America recommend a cut of at least a third in meat consumption to check an epidemic of heart attacks, some grain going now to beef cattle could in principle be diverted.) Similarly fertiliser is absolutely scarce and only if industrialised states cut their consumption a little can the extra 500,000 tons needed now for India's next harvests be made available—the amount is, incidentally, less than the affluent nations use on lawns and golf courses. The problem nagging at the back of many minds at the assembly was whether this condition of absolute shortage is strictly temporary or the first premonition of a profounder change.

The truth is that no one felt they knew. As Dr Scheel put it: "Unreliable data, accelerated changes, the impossibility of foreseeing developments—this is where governments and countries come up against their limitations". The temptation in these conditions is to save oneself and batten down the hatches. In the short run, the raw materials producers are tempted, regardless of the outcome, to seek much higher prices by Opec-type cartels and to use their pre-eminence in numbers to reverse the dominance of established wealth. Equally, the already rich could cut aid, protect their reserves and their industries, try for self-sufficiency and turn their backs on the troublesome poor. Yet in spite of the language, the surface behaviour and much of the rhetoric, this did not appear to be happening at the special session.

A majority of the delegations were ready for dialogue, searching for greater leadership, obscurely aware of interdependence and deeply afraid of some precipitate catastrophe. The opportunity thus offered to the leaders of the United Nations system is alarmingly great. The fact that they have been offered it suggests that even the most powerful communities are beginning to wonder whether they can go it alone.

DECLARATION ON THE ESTABLISHMENT OF A NEW INTERNATIONAL ECONOMIC ORDER

We, the Members of the United Nations,
Having convened a special session of the General Assembly to study for the first time the problems of raw materials and development, devoted to the consideration of the most important economic problems facing the world community,

Bearing in mind the spirit, purposes and principles of the Charter of the United Nations to promote the economic advancement and social progress of all peoples,

Solemnly proclaim our united determination to work urgently for

THE ESTABLISHMENT OF A NEW INTERNATIONAL ECONOMIC ORDER

based on equity, sovereign equality, interdependence, common interest and co-operation among all States, irrespective of their economic and social systems which shall correct inequalities and redress existing injustices, make it possible to eliminate the widening gap between the developed and the developing countries and ensure steadily accelerating economic and social development and peace and justice for present and future generations, and, to that end, declare:

1. The greatest and most significant achievement during the last decades has been the independence from colonial and alien domination of a large number of peoples and nations which has enabled them to become members of the community of free peoples. Technological progress has also been made in all spheres of economic activities in the last three decades, thus providing a solid potential for improving the well-being of all peoples. However, the remaining vestiges of alien and colonial domination, foreign occupation, racial discrimination, *apartheid* and neo-colonialism in all its forms continue to be among the greatest obstacles to the full emancipation and progress of the develop-

ing countries and all the peoples involved. The benefits of technological progress are not shared equitably by all members of the international community. The developing countries, which constitute 70 per cent of the world's population, account for only 30 per cent of the world's income. It has proved impossible to achieve an even and balanced development of the international community under the existing international economic order. The gap between the developed and the developing countries continues to widen in a system which was established at a time when most of the developing countries did not even exist as independent States and which perpetuates inequality.

2. The present international economic order is in direct conflict with current developments in international political and economic relations. Since 1970, the world economy has experienced a series of grave crises which have had severe repercussions, especially on the developing countries because of their generally greater vulnerability to external economic impulses. The developing world has become a powerful factor that makes its influence felt in all fields of international activity. These irreversible changes in the relationship of forces in the world necessitate the active, full and equal participation of the developing countries in the formulation and application of all decisions that concern the international community.

3. All these changes have thrust into prominence the reality of interdependence of all the members of the world community. Current events have brought into sharp focus the realization that the interests of the developed countries and those of the developing countries can no longer be isolated from each other, that there is close interrelationship between the prosperity of the developed countries and the growth and development of the developing countries, and that the prosperity of the international community as a whole depends upon the prosperity of its constituent parts. International co-operation for development is the shared goal and common duty of all countries. Thus the political, economic and social well-being of present and future generations depends more than ever on co-operation between all members of the international community on the basis of sovereign equality and the removal of the disequilibrium that exists between them.

4. The new international economic order should be founded on full respect for the following principles:

(a) Sovereign equality of States, self-determination of all peoples, inadmissibility of the acquisition of territories by force, territorial integrity and non-interference in the internal affairs of other States;

(b) The broadest co-operation of all the States members of the international community, based on equity, whereby the prevailing disparities in the world may be banished and prosperity secured for all;

(c) Full and effective participation on the basis of equality of all countries in the solving of world economic problems in the common interest of all countries, bearing in mind the necessity to ensure the accelerated development of all the developing countries, while devoting particular attention to the adoption of special measures in favour of the least developed, land-locked and island developing countries as well as those developing countries most seriously affected by economic crises and natural calamities, without losing sight of the interests of other developing countries;

(d) The right to every country to adopt the economic and social system that it deems to be the most appropriate for its own development and not to be subjected to discrimination of any kind as a result;

(e) Full permanent sovereignty of every State over its natural resources and all economic activities. In order to safeguard these resources, each State is entitled to exercise effective control over them and their exploitation with means suitable to its own situation, including the right to nationalization or transfer of ownership to its nationals, this right being an expression of the full permanent sovereignty of the State. No State may be subjected to economic, political or any other type of coercion to prevent the free and full exercise of this inalienable right;

(f) The right of all States, territories and peoples under foreign occupation, alien and colonial domination or *apartheid* to restitution and full compensation for the exploitation and depletion of, and damages to, the natural resources and all other resources of those States, territories and peoples;

(g) Regulation and supervision of the activities of transnational corporations by taking measures in the interest of the national economies of the countries where such transnational corporations operate on the basis of the full sovereignty of those countries;

(h) The right of the developing countries and the peoples of territories under colonial and racial domination and foreign occupation to achieve their liberation and to regain effective control over their natural resources and economic activities;

(i) The extending of assistance to developing countries, peoples and territories which are under colonial and alien domination, foreign occupation, racial discrimination or *apartheid* or are subjected to economic, political or any other type of coercive measures to obtain from them the subordination of the exercise of their sovereign rights and to secure from them advantages of any kind, and to neo-colonialism in all its forms, and which have established or are endeavouring to establish effective control over their natural resources and economic activities that have been or are still under foreign control;

(*j*)   Just and equitable relationship between the prices of raw materials, primary products, manufactured and semi-manufactured goods exported by developing countries and the prices of raw materials, primary commodities, manufactures, capital goods and equipment imported by them with the aim of bringing about sustained improvement in their unsatisfactory terms of trade and the expansion of the world economy;

(*k*)   Extension of active assistance to developing countries by the whole international community, free of any political or military conditions;

(*l*)   Ensuring that one of the main aims of the reformed international monetary system shall be the promotion of the development of the developing countries and the adequate flow of real resources to them;

(*m*)   Improving the competitiveness of natural materials facing competition from synthetic substitutes;

(*n*)   Preferential and non-reciprocal treatment for developing countries, wherever feasible, in all fields of international economic co-operation whenever possible;

(*o*)   Securing favourable conditions for the transfer of financial resources to developing countries;

(*p*)   Giving to the developing countries access to the achievements of modern science and technology, and promoting the transfer of technology and the creation of indigenous technology for the benefit of the developing countries in forms and in accordance with procedures which are suited to their economies;

(*q*)   The need for all States to put an end to the waste of natural resources, including food products;

(*r*)   The need for developing countries to concentrate all their resources for the cause of development;

(*s*)   The strengthening, through individual and collective actions, of mutual economic, trade, financial and technical co-operation among the developing countries, mainly on a preferential basis;

(*t*)   Facilitating the role which producers' associations may play within the framework of international co-operation and, in pursuance of their aims, *inter alia* assisting in the promotion of sustained growth of world economy and accelerating the development of developing countries.

5. The unanimous adoption of the International Development Strategy for the Second United Nations Development Decade was an impor-

tant step in the promotion of international economic co-operation on a just and equitable basis. The accelerated implementation of obligations and commitments assumed by the international community within the framework of the Strategy, particularly those concerning imperative development needs of developing countries, would contribute significantly to the fulfilment of the aims and objectives of the present Declaration.

6. The United Nations as a universal organization should be capable of dealing with problems of international economic co-operation in a comprehensive manner and ensuring equally the interests of all countries. It must have an even greater role in the establishment of a new international economic order. The Charter of Economic Rights and Duties of States, for the preparation of which the present Declaration will provide an additional source of inspiration, will constitute a significant contribution in this respect. All the States Members of the United Nations are therefore called upon to exert maximum efforts with a view to securing the implementation of the present Declaration, which is one of the principal guarantees for the creation of better conditions for all peoples to reach a life worthy of human dignity.

7. The present Declaration on the Establishment of a New International Economic Order shall be one of the most important bases of economic relations between all peoples and all nations.

2229th plenary meeting
1 May 1974

PROGRAMME OF ACTION ON THE ESTABLISHMENT OF A NEW INTERNATIONAL ECONOMIC ORDER

Introduction

1. In view of the continuing severe economic imbalance in the relations between developed and developing countries, and in the context of the constant and continuing aggravation of the imbalance of the economies of the developing countries and the consequent need for the mitigation of their current economic difficulties, urgent and effective measures need to be taken by the international community to assist the developing countries, while devoting particular attention to the least developed, land-locked and island developing countries and those developing countries most seriously affected by economic crises and natural calamities leading to serious retardation of development processes.

2. With a view to ensuring the application of the Declaration on the Establishment of a New International Economic Order, it will be necessary to adopt and implement within a specified period a programme of action of unprecedented scope and to bring about maximum economic co-operation and understanding among all States, particularly between developed and developing countries, based on the principles of dignity and sovereign equality.

I   Fundamental problems of raw materials and primary commodities as related to trade and development

1.   Raw materials

All efforts should be made:

(a)   To put an end to all forms of foreign occupation, racial discrimination, *apartheid,* colonial, neo-colonial and alien domination and exploitation through the exercise of permanent sovereignty over natural resources;

(b)   To take measures for the recovery, exploitation, development, marketing and distribution of natural resources, particularly of developing countries, to serve their national interests, to promote collective self-reliance among them and to strengthen mutually beneficial international economic co-operation with a view to bringing about the accelerated development of developing countries;

(c)   To facilitate the functioning and to further the aims of producers' associations, including their joint marketing arrangements, orderly commodity trading, improvement in export income of producing developing countries and in their terms of trade, and sustained growth of the world economy for the benefit of all;

(d)   To evolve a just and equitable relationship between the prices of raw materials, primary commodities, manufactured and semi-manufactured goods exported by developing countries and the prices of raw materials, primary commodities, food, manufactured and semi-manufactured goods and capital equipment imported by them, and to work for a link between the prices of exports of developing countries and the prices of their imports from developed countries;

(e)   To take measures to reverse the continued trend of stagnation or decline in the real price of several commodities exported by developing countries, despite a general rise in commodity prices, resulting in a decline in the export earnings of these developing countries;

(f)   To take measures to expand the markets for natural products in relation to synthetics, taking into account the interests of the developing countries, and to utilize fully the ecological advantages of these products;

(g)   To take measures to promote the processing of raw materials in the producer developing countries.

2.   **Food**

All efforts should be made:

(a)   To take full account of specific problems of developing countries, particularly in times of food shortages, in the international efforts connected with the food problem;

(b)   To take into account that, owing to lack of means, some developing countries have vast potentialities of unexploited or underexploited land which, if reclaimed and put into practical use, would contribute considerably to the solution of the food crisis;

(c) By the international community to undertake concrete and speedy measures with a view to arresting desertification, salination and damage by locusts or any other similar phenomenon involving several developing countries, particularly in Africa, and gravely affecting the agricultural production capacity of these countries, and also to assist the developing countries affected by this phenomenon to develop the affected zones with a view to contributing to the solution of their food problems;

(d) To refrain from damaging or deteriorating natural resources and food resources, especially those derived from the sea, by preventing pollution and taking appropriate steps to protect and reconstitute those resources;

(e) By developed countries, in evolving their policies relating to production, stocks, imports and exports of food, to take full account of the interests of:

   (i) Developing importing countries which cannot afford high prices for their imports;

   (ii) Developing exporting countries which need increased market opportunities for their exports;

(f) To ensure that developing countries can import the necessary quantity of food without undue strain on their foreign exchange resources and without unpredictable deterioration in their balance of payments, and, in this context, that special measures are taken in respect of the least developed, the landlocked and island developing countries as well as those developing countries most seriously affected by economic crises and natural calamities;

(g) To ensure that concrete measures to increase food production and storage facilities in developing countries are introduced, *inter alia,* by ensuring an increase in all available essential inputs, including fertilizers, from developed countries on favourable terms;

(h) To promote exports of food products of developing countries through just and equitable arrangements, *inter alia,* by the progressive elimination of such protective and other measures as constitute unfair competition.

3. **General trade**

All efforts should be made:

(a) To take the following measures for the amelioration of terms of trade of developing countries and concrete steps to eliminate chronic trade deficits of developing countries;

(i) Fulfilment of relevant commitments already undertaken in the United Nations Conference on Trade and Development and in the International Development Strategy for the Second United Nations Development Decade;

(ii) Improved access to markets in developed countries through the progressive removal of tariff and non-tariff barriers and of restrictive business practices;

(iii) Expeditious formulation of commodity agreements where appropriate, in order to regulate as necessary and to stabilize the world markets for raw materials and primary commodities;

(iv) Preparation of an over-all integrated programme, setting out guidelines and taking into account the current work in this field, for a comprehensive range of commodities of export interest to developing countries;

(v) Where products of developing countries compete with the domestic production in developed countries, each developed country should facilitate the expansion of imports from developing countries and provide a fair and reasonable opportunity to the developing countries to share in the growth of the market;

(vi) When the importing developed countries derive receipts from customs duties, taxes and other protective measures applied to imports of these products, consideration should be given to the claim of the developing countries that these receipts should be reimbursed in full to the exporting developing countries or devoted to providing additional resources to meet their development needs;

(vii) Developed countries should make appropriate adjustments in their economies so as to facilitate the expansion and diversification of imports from developing countries and thereby permit a rational, just and equitable international division of labour;

(viii) Setting up general principles for pricing policy for exports of commodities of developing countries, with a view to rectifying and achieving satisfactory terms of trade for them;

(ix) Until satisfactory terms of trade are achieved for all developing countries, consideration should be given to alternative means, including improved compensatory financing schemes for meeting the development needs of the developing countries concerned;

(x) Implementation, improvement and enlargement of the generalized system of preferences for exports of agricultural primary commodities, manufactures and semi-manufactures from developing to developed countries and consideration of its extension to commodities, including those which are processed or semi-processed; developing countries which are or will be sharing their existing tariff advantages in some developed countries as the result of the introduction and eventual enlargement of the generalized system of preferences should, as a matter of urgency, be granted new openings in the markets of other developed countries which should offer them export opportunities that at least compensate for the sharing of those advantages;

(xi) The setting up of buffer stocks within the framework of commodity arrangements and their financing by international financial institutions, wherever necessary, by the developed countries and, when they are able to do so, by the developing countries, with the aim of favouring the producer developing and consumer developing countries and of contributing to the expansion of world trade as a whole;

(xii) In cases where natural materials can satisfy the requirements of the market, new investment for the expansion of the capacity to produce synthetic materials and substitutes should not be made.

(b) To be guided by the principles of non-reciprocity and preferential treatment of developing countries in multilateral trade negotiations between developed and developing countries, and to seek sustained and additional benefits for the international trade of developing countries, so as to achieve a substantial increase in their foreign exchange earnings, diversification of their exports and acceleration of the rate of their economic growth.

4. **Transportation and insurance**

All efforts should be made:

(a) To promote an increasing and equitable participation of developing countries in the world shipping tonnage;

(b) To arrest and reduce the ever-increasing freight rates in order to reduce the cost of imports to, and exports from, the developing countries;

(c) To minimize the cost of insurance and reinsurance for developing countries and to assist the growth of domestic insurance and reinsurance markets in developing countries and the establishment to this end, where appropriate, of institutions in these countries or at the regional level;

(d) To ensure the early implementation of the code of conduct for liner conferences;

(e) To take urgent measures to increase the import and export capability of the least developed countries and to offset the disadvantages of the adverse geographic situation of land-locked countries, particularly with regard to their transportation and transit costs, as well as developing island countries in order to increase their trading ability;

(f) By the developed countries to refrain from imposing measures or implementing policies designed to prevent the importation, at equitable prices, of commodities from the developing countries or from frustrating the implementation of legitimate measures and policies adopted by the developing countries in order to improve prices and encourage the export of such commodities.

II International monetary system and financing of the development of developing countries

1. Objectives

All efforts should be made to reform the international monetary system with, *inter alia,* the following objectives:

(a) Measures to check the inflation already experienced by the developed countries, to prevent it from being transferred to developing countries and to study and devise possible arrangements within the International Monetary Fund to mitigate the effects of inflation in developed countries on the economies of developing countries;

(b) Measures to eliminate the instability of the international monetary system, in particular the uncertainty of the exchange rates, especially as it affects adversely the trade in commodities;

(c) Maintenance of the real value of the currency reserves of the developing countries by preventing their erosion from inflation and exchange rate depreciation of reserve currencies;

(d) Full and effective participation of developing countries in all phases of decision-making for the formulation of an equitable and durable monetary system and adequate participation of developing countries in all bodies entrusted with this reform and, particularly, in the Board of Governors of the International Monetary Fund;

(e) Adequate and orderly creation of additional liquidity with particular regard to the needs of the developing countries through the additional allocation of special drawing rights based on the concept of world liquidity needs to be appropriately revised in the light of the new international environment; any creation of international liquidity should be made through international multilateral mechanisms;

(f) Early establishment of a link between special drawing rights and additional development financing in the interest of developing countries, consistent with the monetary characteristics of special drawing rights;

(g) Review by the International Monetary Fund of the relevant provisions in order to ensure effective participation by developing countries in the decision-making process;

(h) Arrangements to promote an increasing net transfer of real resources from the developed to the developing countries;

(i) Review of the methods of operation of the International Monetary Fund, in particular the terms for both credit repayments and "stand-by" arrangements, the system of compensatory financing, and the terms of the financing of commodity buffer stocks, so as to enable the developing countries to make more effective use of them.

2.   Measures

All efforts should be made to take the following urgent measures to finance the development of developing countries and to meet the balance-of-payment crises in the developing world:

(a) Implementation at an accelerated pace by the developed countries of the time-bound programme, as already laid down in the International Development Strategy for the Second United Nations Development Decade, for the net amount of financial resource transfers to developing countries; increase in the official component of the net amount of financial resource transfers to developing countries so as to meet and even to exceed the target of the Strategy;

(b) International financing institutions should effectively play their role as development financing banks without discrimination on account of the political or economic system of any member country, assistance being untied;

(c) More effective participation by developing countries, whether recipients or contributors, in the decision-making process in the competent organs of the International Bank for Reconstruction and Development and the International Development Association, through the establishment of a more equitable pattern of voting rights;

(*d*) Exemption, wherever possible, of the developing countries from all import and capital outflow controls imposed by the developed countries;

(*e*) Promotion of foreign investment, both public and private, from developed to developing countries in accordance with the needs and requirements in sectors of their economies as determined by the recipient countries;

(*f*) Appropriate urgent measures, including international action, should be taken to mitigate adverse consequences for the current and future development of developing countries arising from the burden of external debt contracted on hard terms;

(*g*) Debt renegotiation on a case-by-case basis with a view to concluding agreements on debt cancellation, moratorium, rescheduling or interest subsidization;

(*h*) International financial institutions should take into account the special situation of each developing country in reorienting their lending policies to suit these urgent needs; there is also need for improvement in practices of international financial institutions in regard to, *inter alia,* development financing and international monetary problems;

(*i*) Appropriate steps should be taken to give priority to the least developed, land-locked and island developing countries and to the countries most seriously affected by economic crises and natural calamities, in the availability of loans for development purposes which should include more favourable terms and conditions.

III Industrialization

All efforts should be made by the international community to take measures to encourage the industrialization of the developing countries, and, to this end:

(*a*) The developed countries should respond favourably, within the framework of their official aid as well as international financial institutions, to the requests of developing countries for the financing of industrial projects;

(*b*) The developed countries should encourage investors to finance industrial production projects, particularly export-oriented production, in developing countries, in agreement with the latter and within the context of their laws and regulations;

(*c*) With a view to bringing about a new international economic structure which should increase the share of the developing countries in world industrial production, the developed countries and the agencies of the United Nations system, in co-operation with the developing countries, should contribute to setting up new industrial capacities including raw materials and commodity-transforming facilities as a matter of priority in the developing countries that produce those raw materials and commodities;

(d) The international community should continue and expand, with the aid of the developed countries and the international institutions, the operational and instruction-oriented technical assistance programmes, including vocational training and management development of national personnel of the developing countries, in the light of their special development requirements.

IV Transfer of technology

All efforts should be made:

(a) To formulate an international code of conduct for the transfer of technology corresponding to needs and conditions prevalent in developing countries;

(b) To give access on improved terms to modern technology and to adapt that technology, as appropriate, to specific economic, social and ecological conditions and varying stages of development in developing countries;

(c) To expand significantly the assistance from developed to developing countries in research and development programmes and in the creation of suitable indigenous technology;

(d) To adapt commercial practices governing transfer of technology to the requirements of the developing countries and to prevent abuse of the rights of sellers;

(e) To promote international co-operation in research and development in exploration and exploitation, conservation and the legitimate utilization of natural resources and all sources of energy.

In taking the above measures, the special needs of the least developed and land-locked countries should be borne in mind.

V Regulation and control over the activities of transnational corporations

All efforts should be made to formulate, adopt and implement an international code of conduct for transnational corporations:

(a) To prevent interference in the internal affairs of the countries where they operate and their collaboration with racist régimes and colonial administrations;

(b) To regulate their activities in host countries, to eliminate restrictive business practices and to conform to the national development plans and objectives of developing countries, and in this context facilitate, as necessary, the review and revision of previously concluded arrangements;

(c) To bring about assistance, transfer of technology and management skills to developing countries on equitable and favourable terms;

(d) To regulate the repatriation of the profits accruing from their operations, taking into account the legitimate interests of all parties concerned;

(e) To promote reinvestment of their profits in developing countries.

VI Charter of Economic Rights and Duties of States

The Charter of Economic Rights and Duties of States, the draft of which is being prepared by a working group of the United Nations and which the General Assembly has already expressed the intention of adopting at its twenty-ninth regular session, shall constitute an effective instrument towards the establishment of a new system of international economic relations based on equity, sovereign equality, and interdependence of the interests of developed and developing countries. It is therefore of vital importance that the aforementioned Charter be adopted by the General Assembly at its twenty-ninth session.

VII Promotion of co-operation among developing countries

1. Collective self-reliance and growing co-operation among developing countries will further strengthen their role in the new international economic order. Developing countries, with a view to expanding co-operation at the regional, subregional and interregional levels, should take further steps, *inter alia*:

(a) To support the establishment and/or improvement of an appropriate mechanism to defend the prices of their exportable commodities and to improve access to and stabilize markets for them. In this context the increasingly effective mobilization by the whole group of oil-exporting countries of their natural resources for the benefit of their economic development is to be welcomed. At the same time there is the paramount need for co-operation among the developing countries in evolving urgently and in a spirit of solidarity all possible means to assist developing countries to cope with the immediate problems resulting from this legitimate and perfectly justified action. The measures already taken in this regard are a positive indication of the evolving co-operation between developing countries;

(b)   To protect their inalienable right to permanent sovereignty over their natural resources;

(c)   To promote, establish or strengthen economic integration at the regional and subregional levels;

(d)   To increase considerably their imports from other developing countries;

(e)   To ensure that no developing country accords to imports from developed countries more favourable treatment than that accorded to imports from developing countries. Taking into account the existing international agreements, current limitations and possibilities and also their future evolution, preferential treatment should be given to the procurement of import requirements from other developing countries. Wherever possible, preferential treatment should be given to imports from developing countries and the exports of those countries;

(f)   To promote close co-operation in the fields of finance, credit relations and monetary issues, including the development of credit relations on a preferential basis and on favourable terms;

(g)   To strengthen efforts which are already being made by developing countries to utilize available financial resources for financing development in the developing countries through investment, financing of export-oriented and emergency projects and other long-term assistance;

(h)   To promote and establish effective instruments of co-operation in the fields of industry, science and technology, transport, shipping and mass communication media.

2.   Developed countries should support initiatives in the regional, subregional and interregional co-operation of developing countries through the extension of financial and technical assistance by more effective and concrete actions, particularly in the field of commercial policy.

VIII Assistance in the exercise of permanent sovereignty of States over natural resources

All efforts should be made:

(a)   To defeat attempts to prevent the free and effective exercise of the rights of every State to full and permanent sovereignty over its natural resources;

(b)   To ensure that competent agencies of the United Nations system meet requests for assistance from developing countries in connexion with the operation of nationalized means of production.

IX   Strengthening the role of the United Nations system in the field of international economic co-operation

1.   In furtherance of the objectives of the International Development Strategy for the Second United Nations Development Decade and in accordance with the aims and objectives of the Declaration on the Establishment of a New International Economic Order, all Member States pledge to make full use of the United Nations system in the implementation of the present Programme of Action, jointly adopted by them, in working for the establishment of a new international economic order and thereby strengthening the role of the United Nations in the field of world-wide co-operation for economic and social development.

2.   The General Assembly of the United Nations shall conduct an over-all review of the implementation of the Programme of Action as a priority item. All the activities of the United Nations system to be undertaken under the Programme of Action as well as those already planned, such as the World Population Conference, 1974, the World Food Conference, the Second General Conference of the United Nations Industrial Development Organization and the mid-term review and appraisal of the International Development Strategy for the Second United Nations Development Decade should be so directed as to enable the special session of the General Assembly on development, called for under Assembly resolution 3172 (XXVIII) of 17 December 1973, to make its full contribution to the establishment of the new international economic order. All Member States are urged, jointly and individually, to direct their efforts and policies towards the success of that special session.

3.   The Economic and Social Council shall define the policy framework and co-ordinate the activities of all organizations, institutions and subsidiary bodies within the United Nations system which shall be entrusted with the task of implementing the present Programme of Action. In order to enable the Economic and Social Council to carry out its tasks effectively:

   (a)   All organizations, institutions and subsidiary bodies concerned within the United Nations system shall submit to the Economic and Social Council progress reports on the implementation of the Programme of Action within their respective fields of competence as often as necessary, but not less than once a year;

   (b)   The Economic and Social Council shall examine the progress reports as a matter of urgency, to which end it may be convened, as necessary, in special session or, if need be, may function continuously. It shall draw the attention of the General Assembly to the problems and difficulties arising in connexion with the implementation of the Programme of Action.

4. All organizations, institutions, subsidiary bodies and conferences of the United Nations system are entrusted with the implementation of the Programme of Action. The activities of the United Nations Conference on Trade and Development, as set forth in General Assembly resolution 1995 (XIX) of 30 December 1964, should be strengthened for the purpose of following in collaboration with other competent organizations the development of international trade in raw materials throughout the world.

5. Urgent and effective measures should be taken to review the lending policies of international financial institutions, taking into account the special situation of each developing country, to suit urgent needs, to improve the practices of these institutions in regard to, *inter alia*, development financing and international monetary problems, and to ensure more effective participation by developing countries — whether recipients or contributors — in the decision-making process through appropriate revision of the pattern of voting rights.

6. The developed countries and others in a position to do so should contribute substantially to the various organizations, programmes and funds established within the United Nations system for the purpose of accelerating economic and social development in developing countries.

7. The present Programme of Action complements and strengthens the goals and objectives embodied in the International Development Strategy for the Second United Nations Development Decade as well as the new measures formulated by the General Assembly at its twenty-eighth session to offset the short-falls in achieving those goals and objectives.

8. The implementation of the Programme of Action should be taken into account at the time of the mid-term review and appraisal of the International Development Strategy for the Second United Nations Development Decade. New commitments, changes, additions and adaptations in the Strategy should be made, as appropriate, taking into account the Declaration on the Establishment of a New International Economic Order and the present Programme of Action.

X   Special Programme

The General Assembly adopts the following Special Programme, including particularly emergency measures to mitigate the difficulties of the developing countries most seriously affected by economic crisis, bearing in mind the particular problem of the least developed and land-locked countries:

The General Assembly,

Taking into account the following considerations:

(a)   The sharp increase in the prices of their essential imports such as food, fertilizers, energy products, capital goods, equipment and services, including transportation and transit costs, has gravely exacerbated the increasingly adverse terms of trade of a number of developing countries, added to the burden of their foreign debt and, cumulatively, created a situation which, if left untended, will make it impossible for them to finance their essential imports and development and result in a further deterioration in the levels and conditions of life in these countries. The present crisis is the outcome of all the problems that have accumulated over the years: in the field of trade, in monetary reform, the world-wide inflationary situation, inadequacy and delay in provision of financial assistance and many other similar problems in the economic and developmental fields. In facing the crisis, this complex situation must be borne in mind so as to ensure that the Special Programme adopted by the international community provides emergency relief and timely assistance to the most seriously affected countries. Simultaneously, steps are being taken to resolve these outstanding problems through a fundamental restructuring of the world economic system, in order to allow these countries while solving the present difficulties to reach an acceptable level of development.

(b)   The special measures adopted to assist the most seriously affected countries must encompass not only the relief which they require on an emergency basis to maintain their import requirements, but also, beyond that, steps to consciously promote the capacity of these countries to produce and earn more. Unless such a comprehensive approach is adopted, there is every likelihood that the difficulties of the most seriously affected countries may be perpetuated. Nevertheless, the first and most pressing task of the international community is to enable these countries to meet the short-fall in their balance-of-payments positions. But this must be simultaneously supplemented by additional development assistance to maintain and thereafter accelerate their rate of economic development.

(c)   The countries which have been most seriously affected are precisely those which are at the greatest disadvantage in the world economy: the least developed, the land-locked and other low-income developing countries as well as other developing countries whose economies have been seriously dislocated as a result of the present economic crisis, natural calamities, and foreign aggression and occupation. An indication of the countries thus affected, the level of the impact on their economies and the kind of relief and assistance they require can be assessed on the basis, *inter alia,* of the following criteria:

   (i)   Low *per capita* income as a reflection of relative poverty, low productivity, low level of technology and development;

   (ii) Sharp increase in their import cost of essentials relative to export earnings;

   (iii) High ratio of debt servicing to export earnings;

   (iv) Insufficiency in export earnings, comparative inelasticity of export incomes and unavailability of exportable surplus;

   (v)   Low level of foreign exchange reserves or their inadequacy for requirements;

   (vi) Adverse impact of higher transportation and transit costs;

   (vii) Relative importance of foreign trade in the development process.

(d)   The assessment of the extent and nature of the impact on the economies of the most seriously affected countries must be made flexible, keeping in mind the present uncertainty in the world economy, the adjustment policies that may be adopted by the developed countries and the flow of capital and investment. Estimates of the payments situation and needs of these countries can be assessed and projected reliably only on the basis of their average performance over a number of years. Long-term projections, at this time, cannot but be uncertain.

(e)   It is important that, in the special measures to mitigate the difficulties of the most seriously affected countries, all the developed countries as well as the developing countries should contribute according to their level of development and the capacity and strength of their economies. It is notable that some developing countries, despite their own difficulties and development needs, have shown a willingness to play a concrete and helpful role in ameliorating the difficulties faced by the poorer developing countries. The various initiatives and measures taken recently by certain developing countries with adequate resources on a bilateral and multilateral basis to contribute to alleviating the difficulties of other developing countries are a reflection of their commitment to the principle of effective economic co-operation among developing countries.

(f) The response of the developed countries which have by far the greater capacity to assist the affected countries in overcoming their present difficulties must be commensurate with their responsibilities. Their assistance should be in addition to the presently available levels of aid. They shoulf fulfil and if possible exceed the targets of the International Development Strategy for the Second United Nations Development Decade on financial assistance to the developing countries, especially that relating to official development assistance. They should also give serious consideration to the cancellation of the external debts of the most seriously affected countries. This would provide the simplest and quickest relief to the affected countries. Favourable consideration should also be given to debt moratorium and rescheduling. The current situation should not lead the industrialized countries to adopt what will ultimately prove to be a self-defeating policy aggravating the present crisis.

Recalling the constructive proposals made by His Imperial Majesty the Shahanshah of Iran and His Excellency Mr Houari Boumediène, President of the People's Democratic Republic of Algeria,

1. *Decides* to launch a Special Programme to provide emergency relief and development assistance to the developing countries most seriously affected, as a matter of urgency, and for the period of time necessary, at least until the end of the Second United Nations Development Decade, to help them overcome their present difficulties and to achieve self-sustaining economic development;

2. *Decides* as a first step in the Special Programme to request the Secretary-General to launch an emergency operation to provide timely relief to the most seriously affected developing countries, as defined in subparagraph (c) above, with the aim of maintaining unimpaired essential imports for the duration of the coming 12 months and to invite the industrialized countries and other potential contributors to announce their contributions for emergency assistance, or intimate their intention to do so, by 15 June 1974 to be provided through bilateral or multilateral channels, taking into account the commitments and measures of assistance announced or already taken by some countries, and further requests the Secretary-General to report the progress of the emergency operation to the General Assembly at its twenty-ninth session, through the Economic and Social Council at its fifty-seventh session;

3. *Calls upon* the industrialized countries and other potential contributors to extend to the most seriously affected countries immediate relief and assistance which must be of an order of magnitude that is commensurate with the needs of these countries. Such assistance should be in addition to the existing level of aid and provided at a very early date to the maximum possible extent on a grant basis and, where not possible, on soft terms. The disbursement and relevant operational procedures and terms must reflect this exceptional situation. The assistance could be provided either through bilateral or multilateral channels, including such new institutions and facilities that have been or are to be set up. The special measures may include the following:

(a) Special arrangements on particularly favourable terms and conditions including possible subsidies for and assured supplies of essential commodities and goods;

(b) Deferred payments for all or part of imports of essential commodities and goods;

(c) Commodity assistance, including food aid, on a grant basis or deferred payments in local currencies, bearing in mind that this should not adversely affect the exports of developing countries;

(d) Long-term suppliers' credits on easy terms;

(e) Long-term financial assistance on concessionary terms;

(f) Drawings from special International Monetary Fund facilities on concessional terms;

(g) Establishment of a link between the creation of special drawing rights and development assistance, taking into account the additional financial requirements of the most seriously affected countries;

(h) Subsidies, provided bilaterally or multilaterally, for interest on funds available on commercial terms borrowed by the most seriously affected countries;

(i) Debt renegotiation on a case-by-case basis with a view to concluding agreements on debt cancellation, moratorium or rescheduling;

(j) Provision on more favourable terms of capital goods and technical assistance to accelerate the industrialization of the affected countries;

(k) Investment in industrial and development projects on favourable terms;

(l) Subsidizing the additional transit and transport costs, especially of the land-locked countries;

4. *Appeals* to the developed countries to consider favourably the cancellation, moratorium or rescheduling of the debts of the most seriously affected developing countries, on their request, as an important contribution to mitigating the grave and urgent difficulties of these countries;

5. *Decides* to establish a Special Fund under the auspices of the United Nations, through voluntary contributions from industrialized countries and other potential contributors, as a part of the Special Programme, to provide emergency relief and development assistance, which will commence its operations at the latest by 1 January 1975;

6. *Establishes* an *Ad Hoc* Committee on the Special Programme, composed of 36 Member States appointed by the President of the General Assembly, after appropriate consultations, bearing in mind the purposes of the Special Fund and its terms of reference:

   (*a*) To make recommendations, *inter alia,* on the scope, machinery and modes of operation of the Special Fund, taking into account the need for:

   (i) Equitable representation on its governing body;

   (ii) Equitable distribution of its resources;

   (iii) Full utilization of the services and facilities of existing international organizations;

   (iv) The possibility of merging the United Nations Capital Development Fund with the operations of the Special Fund;

   (v) A central monitoring body to oversee the various measures being taken both bilaterally and multilaterally;

   and, to this end, bearing in mind the different ideas and proposals submitted at the sixth special session, including those put forward by Iran and those made at the 2208th plenary meeting, and the comments thereon, and the possibility of utilizing the Special Fund to provide an alternative channel for normal development assistance after the emergency period;

   (*b*) To monitor, pending commencement of the operations of the Special Fund, the various measures being taken both bilaterally and multilaterally to assist the most seriously affected countries;

   (*c*) To prepare, on the basis of information provided by the countries concerned and by appropriate agencies of the United Nations system, a broad assessment of:

   (i) The magnitude of the difficulties facing the most seriously affected countries;

(ii) The kind and quantities of the commodities and goods essentially required by them;

(iii) Their need for financial assistance;

(iv) Their technical assistance requirements, including especially access to technology;

7. *Requests* the Secretary-General of the United Nations, the Secretary-General of the United Nations Conference on Trade and Development, the President of the International Bank for Reconstruction and Development, the Managing Director of the International Monetary Fund, the Administrator of the United Nations Development Programme and the heads of the other competent international organizations to assist the *Ad Hoc* Committee on the Special Programme in performing the functions assigned to it under paragraph 6 above, and to help, as appropriate, in the operations of the Special Fund;

8. *Requests* the International Monetary Fund to expedite decisions on:

(a) The establishment of an extended special facility with a view to enabling the most seriously affected developing countries to participate in it on favourable terms;

(b) The creation of special drawing rights and the early establishment of the link between their allocation and development financing;

(c) The establishment and operation of the proposed new special facility to extend credits and subsidize interest charges on commercial funds borrowed by Member States, bearing in mind the interests of the developing countries and especially the additional financial requirements of the most seriously affected countries;

9. *Requests* the World Bank Group and the International Monetary Fund to place their managerial, financial and technical services at the disposal of Governments contributing to emergency financial relief so as to enable them to assist without delay in channelling funds to the recipients, making such institutional and procedural changes as may be required;

10. *Invites* the United Nations Development Programme to take the necessary steps, particularly at the country level, to respond on an emergency basis to requests for additional assistance which it may be called upon to render within the framework of the Special Programme;

11. *Requests* the *Ad Hoc* Committee on the Special Programme to submit its report and recommendations to the Economic and Social Council at its fifty-seventh session and invites the Council,

on the basis of its consideration of that report, to submit suitable recommendations to the General Assembly at its twenty-ninth session;

12. *Decides* to consider as a matter of high priority at its twenty-ninth session, within the framework of a new international economic order, the question of special measures for the most seriously affected countries.

2229th plenary meeting
1 May 1974

CONFERENCE OF OPEC MEMBER STATES, ALGIERS, 4 TO 6 MARCH 1975

Introduction

One of the key aspects of the development of producer power has been the relationship between the oil producing countries and the less developed countries of the world. The latter have been severely affected by the increases in the price of oil and one might aspect a degree of antagonism on their part towards the member states of OPEC. Such antagonism has not, in general, been in evidence. Quite the contrary, there is now increasing evidence for some kind of economic and political alliance between these groups of countries.

Some heads of state of OPEC member countries have been at pains to insist that, the oil price increases notwithstanding, their countries remain firmly within the third world. They have laid emphasis on the need for the OPEC group to work closely with the less developed countries.

In this connection Algeria has a dual status, as an important oil producer and as a large country with major development problems of its own. Its President, Houari Boumediene, made an important policy statement on relations between OPEC member states and the third world in general on the occasion of a major OPEC conference in Algiers early in 1975.

SPEECH DELIVERED ON MARCH 4, 1975 AT THE OPENING OF THE FIRST CONFERENCE OF THE SOVEREIGNS AND HEADS OF STATE OF THE OPEC MEMBER COUNTRIES

by President Houari Boumediene

Your Majesties, your Excellencies,

Algeria is proud and honoured to welcome you and to host the first meeting of Heads of State of OPEC member countries.

In the name of the Revolutionary Council and the Government of the People's Republic of Algeria, I greet the Sovereigns and Heads of State and the other prominent personalities with us today, and once again I extend to them my warmest welcome.

As OPEC was born, the Algerian people was paying a tribute of blood in defence of its territorial integrity and natural resources

Shortly after the Fourth Conference of the Non-Aligned Nations, which solemnly committed the developing countries to unite to defend, among other things, the price of their raw materials, the members of OPEC jointly took measures which will be a milestone in the worldwide movement by which peoples are recovering control of their natural resources.

At a time when we have the honour of being host to the Heads of State of the Member Countries of OPEC, we cannot but remember that just when Iran, Iraq, Kuwait, Saudi Arabia and Venezuela were deciding to unite their efforts in a joint struggle against the petroleum monopolies, our war of national liberation was reaching the end of its sixth year.

At that time, the colonial power showed that it was resigned to the inevitable restoration of sovereignty to the Algerian people, but had not given the attempt to maintain its hold over the resources of our subsoil; to this end it was preparing plans to split the national territory into two.

In 1960, the year in which OPEC was born, the Algerian people were entering the final phase of a bloody struggle which was to last for two more years, and during which it was to pay a heavy tribute of blood in order to safeguard the integrity of its territory and defend its sovereignty over its natural resources.

For the Algerian people, the historical circumstances in which it witnessed the birth of OPEC and the action it was itself conducting in the continued struggle for national liberation with a view to regaining control of its natural resources, led it quite naturally to play its part, offering support and the fruits of its modest experience to the noble cause which our Organisation is pledged to defend.

By constancy and cohesion OPEC is contributing to the economic future of the Third World

Founded in a situation where the oil monopolies were all-powerful and the isolation of the petroleum exporting countries was dramatic, OPEC has steadily consolidated its position by constantly extending its base and by making its actions increasingly concrete and vigorous.

Undoubtedly, it was the firmness of OPEC and the exemplary cohesion of its members which enabled our Organisation to play a leading part in demonstrating the considerable strength of the Third World and the effectiveness of unions of raw materials producers. In so doing, it has provided a practical illustration of what the developing countries can do in a field to which the Fourth Conference of Non-Aligned Countries gave absolute priority.

In this connection, it would be no exaggeration to say that OPEC has made a great contribution to the steadily expanding role played by the Third World in the world economy. But imperialism does not easily give up its privileges.

The imperialists are trying to blame the world economic crisis on the member countries of OPEC

The verbal violence and open threats proffered in response to the measures of oil price readjustment, and all the schemes evolved to destroy the effects of these measures, are a characteristic manifestation of imperialism and clearly show that, in this fresh phase of our relations with the industrialized countries, we shall have to redouble our vigilance if we are to defend and consolidate the attainments of our past struggles.

The decisions we took as part of the defence of the rightful interests of our peoples came just in time to provide the leaders of the western economies with an excuse which they are using to conceal their inability or their refusal to take the measures needed to cope with a world economic crisis that began long before the adjustment of oil prices and is now reaching the point at which it becomes a real threat to world stability.

Consequently, the peoples of the industrialized countries, obsessed by galloping inflation and the spectre of unemployment and all kinds of disorder, end by attributing their difficulties to the measures taken by our countries, which are presented to them as the main source of their difficulties. And oil is also being blamed for the severe troubles being experienced by the economies of Third World countries.

We all know that such allegations are a conscious deviation from the truth. The international community as a whole has come to a clear conclusion on the underlying causes of the world economic crisis.

At its sixth extraordinary session, the General Assembly of the United Nations devoted itself for the first time in its history to the study of raw materials and development problems; the present situation, it solemnly declared, is the result of problems that have been allowed to accumulate for years, and is characterized as much by a largely unequal distribution of the results of growth and development as by a lack of any genuine economic co-operation between nations.

The initiatives taken by supporters of confrontation are aimed at a return to the status quo ante as regards the exploitation of the resources of the Third World

Your Majesties, your Excellencies,

A year ago international life was dominated by the actions of the Third World countries, particularly the OPEC countries, which assumed historic responsibilities in re-establishing the legitimate rights of their peoples and bringing the whole world to recognize the serious anomalies that are burdening international economic relations. Now, on the contrary, it appears that those who favour a policy of confrontation are playing the leading roles on the political stage of the world.

It is clear that the increasing number of initiatives which they are taking, whether propaganda, threats or diplomatic steps of all kinds, aim at nothing less than a return to the status quo ante in international economic relations.

As far as we, the oil exporting countries of the Third World, are concerned, it is also clear that these initiatives are directed against the fundamental right of our peoples to draw real benefit from their resources, that is to say their right to development and progress.

The vital role of oil in the OPEC countries and in the world economy makes it important in the overall policy of nations

As regards our Organisation, and taking account of the existing situation, it is necessary not merely to co-ordinate the measures we take on oil prices or on a share in the exploitation of oil within our country; we must also endeavour to promote solutions to wider and more serious problems that our Ministers have had to raise with increasing frequency at the highest level.

The fact is that the vital importance of oil for our peoples and its leading place in the world economy and in our relations with the other peoples of the world means that the decisions we take in connection with our oil are of immediate importance at the level of the overall policy of nations.

Therefore, the seriousness of what is at stake, the scale of the problems and the importance of the measures to be considered have made it imperative for us to meet and take counsel together at the highest level, thus enabling us to reflect quietly together, in a particularly confused and troubled international context, and to assume fully both what stems from our rights and what stems from our duties.

The OPEC countries must regain the initiative on the international stage if they are to defend their achievements, thwart the policy of confrontation and help solve the major problems of the world

Your Majesties, your Excellencies,

Just as the past struggles for the right to decide prices and to control our resources have allowed us to show our ability to ensure the triumph of our just cause, so too the present situation is an exceptional opportunity to show the ability of our countries to assume their international responsibilities and display a healthy attitude in their approach to the problems facing other peoples.

The advocates of confrontation are deeply mistaken if they believe they can blunt the combativeness of our peoples in the noble fight for economic emancipation.

By persisting in their will to show others or to convince themselves that intimidation pays, they are just moving further away from the real problems and avoiding the solutions that history will inevitably dictate sooner or later.

As for us, we are answerable to our peoples: we must strongly reassert their fundamental right to dispose of their natural resources, and defend this right by joint action.

As members of the international community, we must contribute to the solution of the major problems of the day, and assume our international responsibilities in conformity with the principles and decisions of the United Nations.

Lastly, it is for us to call world opinion to witness the real meaning of our actions, and the positive part that our countries have played and will continue to play in support of equity, peace and stability in the world.

The propaganda campaigns and the threats against the OPEC countries are aimed at their fundamental rights over their natural resources

Your Majesties, your Excellencies,

The United Nations have solemnly proclaimed the permanent sovereignty of states over their resources, and the developed countries have explicitly recognized this sovereignty. Nevertheless, the persistence of certain attitudes suggests that, in the eyes of the developed countries of the west, recognition of the rights of our peoples was not intended to go beyond the mere enunciation of a principle or a purely formal act.

The resurgence and rehabilitation, as and when it suits their book, of some outdated theory or philosophy, such as the one which maintains that our oil is, after all, a source of wealth which belongs to the whole of mankind, brings out clearly the direction in which some of the developed countries are heading. Behind all this lies an attempt to evolve legal and moral arguments that could, in due course, be used to justify tomorrow's aggression in the eyes of the world and ultimately to provide an apparently legal or even humanitarian basis for abolishing the right of our peoples to dispose of what belongs to them and has been recovered by drive, struggle and sacrifice.

The OPEC countries must cling with renewed vigour to the United Nations principles and proclaim their determination to defend their sovereignty and their fundamental rights over their natural resources

This means that the right of the peoples to dispose of their resources, which we look on as part of the natural order of things, is not always considered to be automatic when the resources of the Third World countries are concerned. At a time when the use of our oil is one of the main questions being debated at the international level, our countries must cling with renewed vigour to the fundamental principles proclaimed on the subject of the sovereignty of states over their natural resources.

In response to the drive to mobilize world opinion against our countries, we should reassert more strongly than ever our attachment to what is right, and our determination to defend that right, to reject categorically any theory aimed at limiting our ability to exercise this right — in other words, to limit our sovereignty — and to proclaim that, as regards the use of our oil, we will, at all times, be guided by the basic principle that our raw materials are, without limitation or exception, the rightful property of our peoples, and that the exploitation of the materials which we have recovered by struggle and sacrifice, is inconceivable unless the interests of our peoples have been taken fully into consideration.

To be just and acceptable, the price of oil ought to be based on essential factors coupled with the best possible management of the world's energy heritage

As regards prices, our countries do not propose to subscribe to an attitude of selling at knock-down prices, as this would ruin the producer. Nor does it intend to adopt arbitrary and unjustified decisions which would be unduly hard on the consumer.

The position that oil should hold in the world economy — having regard to the existence of other sources of energy — the particular characteristics which distinguish it from its competitors, the part if plays and the part it will play in future in the economies of the producer countries, are all essential factors to which we feel that the fullest consideration should be given if a reliable, equitable and realistic assessment is to be made of the value of oil.

Defining the position our oil should occupy while other sources of energy are also available does not mean turning to a system of using it and of obtaining higher returns on it that would end with one source driving out the other; what it does mean is choosing an appropriate management system for the whole of the world's heritage in the field of energy, and especially deciding on the order of priority for conservation of the various sources in the light of their availability and of their individual characteristics.

As such a system must aim both at drawing optimum benefit from the heritage that can be consumed, and at safeguarding the interests of the peoples, it is understandable that the arguments and attitudes with regard to the choice of system may differ from one country to another, from one area to another, from one people to another.

The oil exporting countries, for instance, in an endeavour to preserve their future, are today finding that they must adopt an approach fundamentally different from that recently followed by the oil companies, and reject any form of exploitation based on the assumption that, of all the energy sources, their oil is the first that will disappear.

Oil is all the more precious in that it is a source of life and a resource that cannot be replaced

Today, it is more than ever essential that this material be preserved.

In view of the very serious food problem which already faces mankind, and which will grow worse, and of the likelihood that the world's capacity to expand certain products subject to natural cycles is nearing its limits, oil will have, through its various by-products, to play a part of prime importance; already it is serving not only as a source of energy, but as a source of life which is all the more precious as it cannot be replaced.

The assessment and management of our energy heritage must take account of the future oil requirements of the producer countries

On the other hand, the assessment and management of the world's energy heritage must not fail to take into account the quantities of oil required by the exporting countries for home consumption and, consequently, of the extent to which this commodity will in future be available on the market.

The present levels of domestic consumption in the exporting countries are absurdly low when compared with their production capacity. To elaborate and apply a policy based on these levels would be tantamount to assuming that the state of underdevelopment of these countries is a fact of life which can never change.

It would indeed be unrealistic to ignore the inevitability of historical change. As a result of such change, the peoples who once were oppressed and ignored are now able to pursue their own aims and satisfy the needs to which they give rise. Nothing could be further from equity and more damaging to the hopes of stability than to go on contesting these changes and endeavouring to prevent them.

The development strategy of the OPEC countries is based on the complete utilization of oil. The security of their future supplies is therefore a cardinal point in their strategy

The ultimate choice which underlies the questions relating to the price of oil is acceptance or rejection of the historical changes which are the real alternatives.

Our governments, whose duty is to watch over our peoples' interests and future, could never agree to remain forever in a situation where oil is nothing more than a commodity to be exported in the crude form as their source of income. One of our main long-term objectives is precisely to change this utterly abnormal arrangement.

Our development strategy rests on the full utilization of oil. Today, it provides the financial means to live, to equip ourselves, to sustain our efforts at industrialization and the build-up of our economy; tomorrow, we will need it if we are to meet our domestic demand, which will increase rapidly, and if we are to sustain the development of the industry we will have created.

Oil will be all the more indispensable since, if we are to achieve real change in our relations with the outside world, it will very soon be the only trump card with which to offset the handicaps that will weigh heavily on the ability of our products to compete on the international markets, owing to the leeway we have to make up in technology, know-how and industrial experience.

Henceforth, it is of strategic importance for us to ensure that in the long term, our economies enjoy absolute security for their oil supplies.

It is in response to the calls from the market that the OPEC countries are producing more than is needed to cover their financial needs

Your Majesties, your Excellencies,

Most of our countries are still exporting enormous quantities of oil, more than is needed to build up the financial resources which they require.

True, we shall satisfy the calls from the energy market, as it will take some time to develop other forms and sources of energy. But at the same time, we must also see that the rhythm of production is continually adapted to the essential requirements of the market in order to meet demand, while safeguarding the interests of our future generations by eliminating wastage and over-exploitation of our reserves. One of the foremost tasks of our Organisation must be the creation of the closest co-ordination and concertation between our countries.

The gradual decline in the OPEC countries' share in world supplies of energy which will inevitably occur as these countries develop will be a welcome contribution to calm international relations

Your Majesties, your Excellencies,

The period in which our reserves are supplying a major part of the world's energy requirements is not, of course, particularly conducive to the calm and mutual understanding that we would like to see in our relations with the developed countries.

Because the life of many other peoples is so critically dependent on our oil, the role that now falls to us as producers is above all an obligation placed on us by history; and which, although it is a privilege, is proving very onerous.

On the other hand, our countries themselves are dependent on their export incomes, and this in turn increases their vulnerability beyond the limits of the acceptable.

Consequently, when the developed countries are one and all starting out on a road which will one day bring them to the point where they need no longer call on our oil, we too must draw closer together and present an unbroken front if we are to free ourselves from the oil market of the industrialized countries. It is the constraints of our strategy and the ineluctable requirements of our economic independence and of our security which dictate the immediate adoption of such an attitude.

Thanks to the progressive introduction of new sources of energy coupled with a further increase in the proportion of our production which will be devoted to our own consumption, the share of the oil exporting countries in world supplies is likely to decline until oil, though it will still be a part of international trade, will revert to a position comparable with that of all other goods.

It is without regret and perhaps with some relief that our countries will see oil, as it were desanctified, lose the present position that makes it the strategic product par excellence and the main subject of discussion in the international arena.

The OPEC countries are in no way imbued with the spirit of revenge. Their main concern is to establish economic stability and cooperation with the developed countries

Today it is common knowledge that the use of our oil and the lives of other peoples are closely linked. We are convinced of the legitimacy of our rights, but for all that we will not shut our eyes to this fact but intend on the contrary, to accept it, for we are by no means motivated by a spirit of revenge that would incite us to penalize peoples, even those of the countries that used to exploit us.

Moreover, what we are doing in our own countries prompts us not to wish for, and still less to foster, the deterioration of the economic situation in the Western countries, but on the contrary to seek together with these countries a profitable cooperation on an ever broader basis.

Are not the social and economic programmes that we have launched in our respective countries (whose achievement depends largely on the cooperation of the developed countries), the large resources we have committed, the challenges and the risks inherent in development which we assume in putting our whole inheritance at stake, are not these the most obvious pledges of our attachment to a stable and harmoniously expanding world economy?

Stability can only be achieved through the efforts of all. It cannot rest on the maintenance of former privileges

It is no less true that this stability must result from the will and endeavours of each and every one of us, for it would be a mere illusion if it were conceived according to the principle of maintaining the privileges acquired in the past.

If the point is to ensure that everyone shall feel secure about his future and the protection of the fruits of his present efforts, then we agree to discuss the problems, to assume our responsibilities and to accept the necessary arrangements which will enable the world to move without setbacks towards a better balance and to greater justice in the world economic order.

If, on the contrary, the intention is to open a discussion that will crudely consist in blaming the oil-exporting countries for all these problems and claiming that these countries should shoulder the whole burden of the measures that will have to be adopted to overcome the existing difficulties, we declare that this is neither fair nor acceptable, and could in any case lead only to deadlock.

Propaganda and threats are false arguments and false solutions which do not intimidate the OPEC countries. To defend their rights, the OPEC countries will use their full solidarity

Your Majesties, your Excellencies,

Peace is undoubtedly the most widely-shared aspiration among all the peoples of the world. For the peoples of the Third World it certainly constitutes an absolute need, since it is a matter of life and death.

Only true international cooperation will permit our peoples to enjoy it fully.

The false arguments and the false solutions which have so far been addressed to us have nothing to do with cooperation. The arrays of figures, statistics and forecasts manipulated solely to accuse us, or even to make us feel guilty, are the kind of false arguments that cannot distort our judgement.

The various attempts to reduce oil consumption, not with the idea of eliminating waste but in the hope of provoking a price fall and weakening us, are false solutions if only because we in reply can reduce our production and increase our prices in order to maintain the level of our incomes.

Indeed, threats directed at us from here and there, and the mobilization of the power of the States that is being undertaken under the cover of the International Energy Agency, certainly worry us but hardly intimidate us, because nothing would be easier than to frustrate any aggression, from whatever quarter, aimed at a foreign takeover of our oil installations.

In reply to those who threaten us, we must proclaim that the OPEC members will stake their total solidarity to defend their rights to their resources.

The International Energy Agency covers an endeavour to prepare solutions based on confrontation

Your Majesties, your Excellencies,

Let us not underrate these threats; the International Energy Agency's openly avowed projects hinge upon the idea of oil scarcity which presupposes total withholding of our exports, and assumption that could be entertained only if the community of OPEC countries is led to give its support to one or more of its members who were victims of an attack aimed at bringing their oil plants under foreign control.

With regard to the reasons advanced by some senior personalities in the developed countries as justification for aggression, world opinion seems to have been focused on the idea of an embargo that would deprive the Western countries of their oil supplies; the possibility of this embargo has been linked with a Zionist aggression against the Arab countries.

In fact, by looking more closely into the statements of certain Western personalities we are led to wonder whether through the idea of strangulation, discreetly substituted for embargo, new types of causus belli are already being predetermined in order to prepare public opinion to accept the possibility of aggression simply in the case of the oil-exporting countries changing the level of production, either by adjusting it to the effective demand or if our countries were led to draw the conclusions from the present monetary policy of the developed countries and consider that, in the last analysis, they would be better advised to keep their oil rather than deliver it against means of payment which are being wilfully and systematically depreciated.

While guarding against the risk of confrontation, the OPEC countries are ready to assume their responsibilities on the international stage

Your Majesties, your Excellencies,

Although it rests with us to guard against any endeavour to provoke confrontation, we must also continue to offer our hand to those developed countries which have chosen cooperation and dialogue.

As a part of the international community, we must be prepared to examine the major problems posed for other countries and be ready to assume our responsibilities: if prices have to be frozen, we will freeze them, if they must be decreased we will decrease them, provided, however, that the developed countries make a similar and simultaneous effort in return, for everyone must contribute, in accordance with his means and responsibilities, to the reorganization of the world economy and the establishment of the stability required for development and prosperity.

The world economic crisis is due to causes within the developed countries and to the fact that there is no real cooperation in support of development

We all know that the world economic crisis is in fact a conjuncture of crucial phases in the life of the nations. In the case of the Third World countries, the state of underdevelopment, inherited from colonial days and aggravated by neo-colonialist exploitation, is being incessantly worsened because international cooperation in support of development is lacking.

With regard to the developed countries, inflation and monetary disorders are the exacerbated expressions of well-known and old phenomena which existed long before the readjustment of the price of petroleum, and which for decades have been finding expression in more or less severe cyclical crises.

The effect of our oil on cost formation in the developed economies was less than one third of one percent before the first price readjustment measures, and is certainly still less than 2% after these measures. These data, which are recorded in all their simplicity in the statistics of the Western countries themselves, show irrefutably that the price of petroleum can have had only secondary effects on the progress of inflation.

Moreover, the freezing of oil prices for the whole of the first half of 1974 in no way prevented the inflation rates in the developed economies from continuing their headlong advance.

Inflation is caused and fed by commercial behaviour in the developed countries and by their desire to live beyond their means

In reality, as everyone is aware, inflation is a phenomenon brought about and sustained by basic commercial behaviour in the developed economies and by the structures and systems of management under which they operate.

Inflation is worsened notably by the fact that the developed countries are living far above their means and that, in order to maintain their standard of living, and particularly the fabulous amounts they spend on arms, prestige investments and substitution programmes for our oil, they have at one and the same time exploited the Third World's resources and created artificial means of payments on a large scale and to such an extent that for years economic disorders have been the rule and the whole international monetary system is today on the verge of collapse.

We do not deny that the adjustment of oil prices has involved an extra burden for the importing industrialized countries.

Without wishing to ignore the problems that face some developed countries, it must be seen that the passing of an era in which oil prices were artificially maintained at unjust and derisory levels has contributed much more to revealing and laying bare existing anomalies in the western economies than to creating really new problems.

An international conference bringing together the developed and the developing countries is to be desired. To be useful and acceptable, it must be representative and deal with everyone's problems

For our part, we are ready to go into these problems at an international conference between industrialized and developing countries. But in order to promote effective, useful and politically acceptable dialogue and co-operation, such a conference must necessarily look not into oil matters alone but also into questions of importance to Third World countries, namely, raw materials and development. It must give equal priority to the major problems facing all parties and must, in particular, lead not only to an easing of the situation of the industrialized countries affected by the crisis but also to elimination of the most urgent difficulties under which the developing countries labour and to setting up means that will enable these countries to start their development.

Lastly, it must be representative, and its participants must reflect the whole international community, even if for reasons of efficiency and convenience it has to be held in a restricted framework.

The OPEC countries must and can contribute to the initiation of constructive cooperation and the elimination of tension

Your Majesties, your Excellencies,

We must accord all the importance it deserves to the gravity of the climate of uncertainty at present prevailing throughout the world: the position of the developing countries most seriously affected by the world economic crisis is worsening, while the United Nations decisions concerning them are blocked; inflation and monetary disorder are daily eating more deeply into the purchasing power of our incomes and the value of our external assets; the most seriously affected developed countries also have reason to worry about their position.

The OPEC members should contribute to defusing the tension straining international relations, to creating a climate of serenity and to initiating a movement of positive and mutually advantageous co-operation. They have the means to do this.

In this connection it should be pointed out that it is time we regained the initiative in order to play a positive part, in an international conference on raw materials and development, and elsewhere.

Algeria has suggested that OPEC should adopt a global proposal on measures to promote international co-operation and development

It was in this same spirit and to this same end that Algeria submitted to the Conference of Ministers of Foreign Affairs, Petroleum and Finance, the draft Overall Proposal on measures to promote development and international co-operation.

The purpose of this Overall Proposal is to invite all countries which have the necessary means to contribute to the immediate adoption of measures that can help solve the major problems of all the countries affected by the crisis. This is why it contains three sets of measures to be applied immediately and jointly, some by OPEC members in support both of the developed countries and of other developing countries, the others by the developed countries in support of the developing countries.

For the industrialized countries, the actions recommended concern oil supplies, price and external disequilibrium.

To help the developed countries, the members of OPEC must make favourable commitments regarding supply, price and the use of the surplus. In return, the developed countries must subscribe to the Overall Proposal

In respect of supply, we recognize the need to allow time for the policies adopted to fight waste and to develop new energy sources to be implemented and bear fruit, and we are prepared to break into the reserves necessary to our future development by exploiting our oilfields beyond the levels strictly required by our financial needs in order to provide the world oil market with the quantities essential for the normal functioning of the world economy.

On prices, it should be recalled that the price we fixed a year ago—well below the level that would result from the mere assessment of the possibilities of substituting other energy sources for our oil, has lost at least a quarter of its purchasing power compared with the beginning of last year.

Nonetheless, we might take into account the period of adjustment which is necessary for the industrialized countries most exposed to the effects of the economic crisis, and renounce any increase of the present price in real terms for a period that might extend to the end of the present decade. Moreover, at the beginning of this period, we could make intermediate arrangements which would allow consumer countries to profit from additional facilities calculated to foster their efforts at economic recovery.

Such concessions, which would mean considerable sacrifices, could be justified only if the industrialized countries bound themselves to subscribe fully to the measures required of them in the Overall Proposal. They also depend on the real effectiveness of the systems to be set up to give due protection to our export incomes and to our external assets.

International co-operation implies that all the parties commit themselves equally. Should the developed countries fail to accept such an approach, we for our part would only be able to resort to the line of action and to the decisions which would enable us to protect the rightful interests of our peoples and to pursue the implementation of our policy of co-operation and solidarity with the other Third World countries.

Use of the external assets of the OPEC countries to help the EEC and other developed countries hit by the crisis must not serve to extend foreign hegemony in these countries

With regard to the balance of payments difficulties affecting some industrialized countries, the search for practical forms of co-operation implies the need to look into the origin of external imbalances and to adjust the help given by our countries in terms of the relative importance of the transitory factors and of the structural causes from which these imbalances arise.

It remains true, however, that the idea of co-operation and mutual aid poses the problem of the acceptance by all concerned of a scale of priorities among world needs and a fair distribution of world resources in accordance with this scale of priorities.

In order to harmonize the objectives, the resources and the worldwide priorities, it is especially important that countries obviously living above their means should accept the fundamental changes required.

For our part, we must then be ready to use our surpluses in such a manner that they usefully serve world economic expansion, particularly by helping redress the external imbalances of the European Economic Community and those of certain other developed countries during the period of adjustment which they need if they are to achieve economic recovery.

It is important that the use of our external assets should be directly negotiated with the European Economic Community members and other developed countries concerned.

At present it is the most powerful who drain the external assets belonging to our countries and then use them as strategic reserves in order to underpin or extend their hegemony over countries facing difficulties and even to obtain from these countries political concessions directed against the Third World countries and especially against the members of OPEC.

This is illustrated by the suggestion made in certain quarters concerning a solidarity fund with conditions of use aimed at nothing less than bringing about a confrontation between the developed countries with external imbalances and the members of OPEC, while those who advocated the creation of this fund intended to establish it largely with our external assets.

Europe, for its part, should not be content to invoke the hegemony with which it is burdened, but try to assert its own political identity and to act in such a manner that the solution to its problems comes initially from its own endeavours and potential.

Sound guarantees must be given to the OPEC countries who hold the surpluses

Moreover, the exporting countries must obtain sound and inalienable guarantees from the developed countries which profit from the use of their external assets.

More than ever, the protection of our external holdings is of prime importance.

All the fuss being made about what are called "petrodollars" in reality expresses a refusal to discard the frame of mind which sanctions the division of the world into two categories of countries, one with all the privileges and rights and destined to direct the world economy, and one condemned to underdevelopment and submission.

The real monetary problems are the result of the anti-democratic functioning of the international monetary system, the unilateral manipulation of the main reserve currencies, arbitrary decisions about prices and the monetary role of gold. These are serious threats which could affect even the security of our countries; it is high time to eliminate them if we sincerely want to establish equity and stability in the world.

It is in the developing countries that we find the most serious and urgent problems. These countries need the means to eliminate their underdevelopment

Your Majesties, your Excellencies,

It is in the developing countries that the most serious problems are found and it is there that the necessary means of overcoming them are most cruelly lacking. Balance-of-payments deficits, for example, reflect situations which have quite different meanings in a developed country and in a Third World country.

For the developed countries, a deficit situation by no means implies a state or even a risk of bankruptcy; it is reflected in most cases by a shortage of liquidity equivalent to a marginal percentage of the Gross National Product. Because of their large assets, their industrial and agricultural patrimony, their infrastructure and production capacities, they have permanently at their disposal the means to fight back and to find remedies; they can slant production either to an increase in their exports or to a reduction in their imports without endangering or even inconveniencing their everyday life; they can also mobilize their external assets, which may be considerable; lastly, they can mobilize stopgap loans on international money markets and more generally all the facilities offered by an international monetary system totally in their hands, especially all the adjustment mechanisms which enable them temporarily to cover their deficit and to have time to use their own resources or redeploy them in the direction required to clear up the crisis situation they are facing. We have seen this in the case of one developed country which is the world's biggest oil importer and which has been able, in less than a year, to redress its trade balance successfully and even to earn a surplus again.

This is not at all the case of Third World countries whose deficits are due to a chronic malady which rapidly assumes dramatic proportions. For, contrary to the developed economies, external trade plays a preponderant part in under-developed economies.

As the industrial production capacities of the developing countries are weak and often non-existent, the satisfaction of their most elementary needs depends on their ability to import; as a result, they are affected even in their everyday life by their deficits, which cannot be offset either by assets that can be mobilized abroad or by any real assistance from the international monetary system, which essentially serves the developed countries.

For these countries, the question is not merely to correct a transitory situation or to provide emergency assistance but to tackle the basic evil, namely underdevelopment. That is why the Sixth Special Session of the United Nations General Assembly was convened.

This Special Session came to decisions which the industrialized countries have so far refused to apply.

The OPEC countries have given tangible proof of their solidarity with the developing countries. They should continue their co-operation with those countries in implementation of the decisions of the United Nations

As members of the community of the Third World, we refuse to make the attitude of non-cooperation adopted by the industrialized countries an excuse for doing nothing on our side and evading our responsibilities.

In this connection a tribute should be paid to all the Third World countries most severely affected by the crisis who have, en bloc, refused to lend themselves to the manœuvres certain powers have been making for the last year to try and set the Third World community against its members who are oil exporters.

In less than one year, our countries have spontaneously taken initiatives by way of financial transfers which, related to our economic and human resources, reach levels never attained by the industrialized countries in helping underdeveloped countries.

The statistics reaching us from IBRD show that in the first eleven months of 1974, our countries as a whole subscribed nearly 17,000 million dollars in credit commitments, or more than 10% of their GNP, whereas actual payments in the same period amounted to 4,000 million dollars, which already represents 2.5% of our countries' total GNP.

All this has been done at very short notice, whereas many of us, the

Arab countries, have every reason to devote all our energies and resources to our own protection, as our security is still threatened by the state of serious tension which persists in our regions by reason of the aggressive machinations of Zionism.

The measures advocated in favour of the developing countries: emergency grants, development credits, credits for purchases of oil, launching of a programme for the manufacture of nitrogenous fertilizers to be supplied to developing countries at production cost, support of raw materials prices

Our efforts on behalf of the other Third World countries should continue within the framework of the solidarity that binds us to them in the following fields:

First: emergency grants to the Third World countries most seriously affected by the economic crisis, as defined in the special programme of the United Nations;

Secondly: allocation of special credits to these countries to enable them to pay for their oil purchases over a period to be determined;

Thirdly: allocation by the oil exporting countries of special credits earmarked for the development of the Third World countries.

These operations can be carried out in a bilateral framework, but to bear more clearly the stamp of international solidarity, they should increasingly be effected at regional or multilateral level.

For this purpose, and in addition to existing channels such as the Arabo-African, Islamic and Latin-American funds, our countries should notably make use of the Special Fund of the United Nations, and may in addition have recourse to the international financial institutions;

Fourthly: immediate launching, within the exporting countries as a whole, of a programme for the manufacture of nitrogenous fertilizers in some ten production units with an output equivalent to two thirds of the oil exporting countries' total current imports of nitrogenous fertilizers; this would all be delivered under the auspices of FAO to the developing countries most affected by the economic crisis, in step with the necessity of developing their agricultural potential. The selling price of these fertilizers should be limited to the cost of production and processing of the raw materials entering into their manufacture; the costs of delivery would also be borne by our countries;

Fifthly: support of operations to sustain at a fair level the prices of raw materials from the Third World countries, should the industrialized countries attempt to force them down.

Pending a genuine reform of the international monetary system, the OPEC countries should have their own instruments for intervention and create a development and international co-operation fund

Your Majesties, your Excellencies,

At a time when many leaders, experts, organizations and institutions are concerning themselves with the use of our external assets, and drawing up, without even bothering to ask what we think about it, various and manifold plans for the use of these assets, would it not be wiser if the OPEC countries' members — for they are, after all the first concerned in the matter — themselves endeavour to make suitable arrangements, under their own direct control, which would enable them, while allowing their assets to fructify, to implement a policy of co-operation with the developing as well as the developed countries?

Algeria suggests that, while waiting for the reform of the international financial institutions to allow them to assume their responsibilities, the the members of OPEC create a fund with a capital of about 10 or 15, 000 million dollars. The fund's resources would be financed out of capital contributions from the members of OPEC and from loans granted by them. Its role would be not only to make the member countries' external assets fructify, but also to channel transfers for co-operative operations, on behalf of the other developing countries and the developed countries, as just described and covered by the Overall Proposal of measures in favour of development and international co-operation, a proposal submitted by Algeria to the Conference of OPEC Member Countries.

Ideas concerning possible ways of managing and operating the fund, and the terms and conditions for its various forms of intervention, are embodied in the Algerian proposals.

The developed countries should for their part commit themselves to concrete and immediate measures in favour of the developing countries

In return for the great efforts we agree to make on their behalf, and in the framework of essential international solidarity, the industrialized countries should, for their part, commit themselves explicitly to a huge undertaking of co-operation towards the development of the Third World countries.

For this purpose, it is essential that their contribution should bring about the speedy elimination of the most urgent problems of the least favoured countries, a quickening in the pace of the developing countries' economic growth, and the strengthening of their ability to change their position and promote their economic social and cultural progress.

The developed countries' contribution should in particular enable the oil exporting countries to achieve their own industrial development and put an end to their dependence on crude oil exports.

In order to be meaningful and effective, this contribution should meet the requirements defined by the decisions of the United Nations and, consequently, take the following lines of action:

First: enable the developing countries to mobilize the totality of their raw materials for the benefit of their own economies, which implies more especially that the selling price for them shall be raised and their real value safeguarded; that the industrialized countries shall cease to obstruct action by producers' unions in the developing countries or accession by the latter to the control of their own economy; that assistance for the processing of raw materials in the countries owning them shall be systematized and institutionalized on an international scale;

Secondly: organize a massive movement of real technological transfers to the developing countries, in order to enable them to manufacture at home the goods they need in every industrial sector, including those involving advanced technologies; this implies that the reluctance so far shown by entrepreneurs in the developed countries shall be entirely put aside, at the instigation of and with a concrete commitment from the governments of their countries;

Thirdly: increase financial transfers in favour of development according to needs and in conformity with the decisions of the United Nations.

The use of capital belonging to oil-exporting countries within the framework of trade credits tied to the supply of equipment and services, having the effect of an immediate return of financial flows towards the supplier developed countries, means that the countries holding such capital shall be given the financial guarantee of the supplier countries, and that borrower developing countries shall accordingly be allowed to raise their ceiling for borrowing;

Fourthly: ban all resort to unilateral decisions in areas affecting the value of the main reserve currencies; this implies the recasting of the international financial institutions so as to abolish the preponderance in them of any one nation and to increase the Third World countries' voting rights, so that they may share in the reform and management of the international monetary system on a footing of strict equality with the developed countries as a whole;

Fifthly: eliminate all discriminatory measures against the oil-exporting countries, in trade and tariff matters and in development assistance.

The pursuit of these lines of action and the practical measures and modes of procedure they imply form the subject of the Overall Proposal on measures in favour of international development and co-operation, which has been submitted by Algeria to the OPEC Ministerial Conference.

Unity and solidarity of the OPEC countries are imperative. They must be centred and amplified in the Third World community

Your Majesties, Your Excellencies,

The magnitude of the challenges we must take up and the importance of what is at stake in the new battles we are to join, in the defence both of our peoples' interests and of our solidarity with the other developing countries, compel us to act in close concert and to promote common action. Our unity and our solidarity of action are the more imperative since we are up against developed countries which have closed their ranks; let us not overestimate their apparent differences, for the reality is that, on the strategic plane, they are in perfect harmony of thought and action.

Clearly, if the developed countries have, despite their potential and their already considerable individual power, deemed it necessary to take concerted action and present a united front, this is even more obviously necessary for our own countries, and nothing would be more perilous for us than to yield to the temptation of acting alone in a context where the regrouping of our forces around common objectives has become the rule.

We can but recognize the obvious truth that each one of us has the greatest possible need to act in the framework of the solidarity which unites not only the members of OPEC but the whole community of developing countries.

It is first of all within this community that we have found, and shall find, the natural support for our actions in defence of our interests, and it is also this community that offers our countries the most immediate and wide prospects of trade and co-operation.

The developed countries' effort to meet development needs will be decisive

Your Majesties, Your Excellencies,

Because colonialist and imperialist exploitation have long deprived them of their wealth and at the same time held them back from progress in science and technology, the developing countries have considerable leeway to make up in the technological, economic, cultural and social fields. Underdevelopment is on the increase, all the more so as the rules and institutions governing our relations with the developed world are systematically unfavourable to our countries, and international co-operation is not forthcoming.

The extent and nature of the needs of the developing countries as a whole are such that the contribution required of the industrialized countries is still decisive, if we are to be given a real chance to succeed in the fight against underdevelopment.

By committing themselves to development, the developed countries will be working for their own economic growth

In agreeing to commit themselves resolutely to co-operate in support of development, the developed countries will not only be answering the dictates of human solidarity and the demand for world stability, they will also be doing something for the prosperity of their peoples. Indeed, by the very fact of embarking on the cumulative process of development, the Third World regions will, because of their size, offer tremendous scope for the expansion of international trade, and so give to all the peoples of the developed countries, new, ever wider and more numerous opportunities for reaping the fruits of their own labours and endeavours, and for increasing their earnings.

The current state of crisis is as loaded with menace as with promise. Will the developed countries be able to seize the opportunity of working towards a mutually profitable co-operation?

In the prevailing world context, both political and economic, where an atmosphere of tension is mixed with a feeling of widespread anxiety, we declare our faith in the possibility of swiftly substituting serenity for anxiety, cooperation for stress.

We are convinced that our appeal, if it were to be heard, and our actions, if the developed countries were willing to join in and sustain them, might usher in a totally new era in the history of international relations.

By shaking the world economy and bringing out into a crude, particularly revealing limelight all the anomalies affecting international economic relations, the current crisis has the undoubted merit of forcing everybody to look at the problems and open his mind to the necessity of major changes in the world economy.

This fact is, by itself, a revolution. It shows that the present crisis, fraught with menace as it is, is also laden with promise.

Will the developed countries succeed in grasping this historic opportunity, and work with us for the achievement of the fundamental aspirations of the peoples of the earth towards equity, freedom from stress and mutually advantageous co-operation?

At all events we, the Member Countries of OPEC, must take positive action to ensure that before the tribunal of history it is clearly established that we have left no stone unturned to give every chance to the realization of this promise.

The initiatives we are able to take today may prove decisive.

For our part, we have no reason to doubt the outcome of our action. Just as history has rewarded, in ways which will be familiar to you, our past struggles and the battles we have waged to recover both our sovereignty and our resources, so too will history reward what we are doing today in defence of our peoples' interests and in co-operation with the other peoples of the world.

THE COMMONWEALTH HEADS OF GOVERNMENT MEETING, KINGSTON, JAMAICA, MAY 1975

Introduction

Most meetings of Commonwealth Heads of Government tend to concentrate on political matters, often of a regional nature such as problems in Southern Africa. The Jamaica meeting was somewhat unusual in that a great deal of time was given to a discussion on the economic relationships between developed and less developed countries. This was largely at the instigation of the British Prime Minister, Harold Wilson.

Mr Wilson had, in a speech in Leeds in February, already drawn attention to the undesirability of the creation of a number of cartels by the third world producers of important raw materials. He considered that such a development would be a force for instability. He prepared a major speech for the Jamaica conference which was supported by detailed documents, the material being published afterwards in London as a White Paper. This unusual step indicated the importance with which Mr Wilson viewed the question of producer power.

His speech and the supporting memorandum are reprinted here and show that his aim was to propose a number of steps to improve the trading prospects of less developed countries whilst ensuring that the developed industrialised countries would be assured of supplies of important raw materials. A series of commodity agreements would be important in this connection.

His proposals were not received with the greatest of enthusiasm by the less developed countries of the Commonwealth, and the speech by Mr Forbes Burnham of Guyana, also reprinted here, is an example of the response. According to Mr Burnham, any move towards stabilising world commodity prices had to be seen as just one part of the programme aimed at achieving a genuine redistribution of world wealth. The much sought-after new international economic order would not be brought about by commodity agreements which should have been approved decades earlier. With the potential of producer power now becoming apparent, third world governments were now interested in much more basic reforms of world economic relationships.

SPEECH BY THE PRIME MINISTER

THE RT HON HAROLD WILSON, OBE, FRS, MP,
AT THE COMMONWEALTH HEADS OF GOVERNMENT MEETING,
KINGSTON, JAMAICA, 1 MAY 1975

All of us here have long recognised the need for economic interdependence in our trade and dealings with one another, and in the wider world. Over generations, failure to make this interdependence a reality has been the cause of great suffering, suffering above all for developing countries producing the food and raw materials the world needs, without a fair and assured return.

If any doubt remained about the need for interdependence surely this has been dispelled by the events of these past few years.

We have all been affected. But by far the hardest hit of all are those developing countries whose pattern of exports denies them any chance of profiting by the boom in commodity prices, while at the same time they have had to pay a lot more for all their essential imports, especially oil and food, fertilisers and feeding stuffs. Most tragic of all has been the effect on nations already facing starvation—starvation aggravated in some cases by drought and others by flood—who have then found their resources strained beyond endurance to pay the increased cost of the things they need.

I want to make it clear in what I propose today that the British Government fully accept that the relationship, the balance, between the rich and poor countries of the world is wrong and must be remedied. That is the principle on which my proposals rest: that the wealth of the world must be redistributed in favour of the poverty stricken and the starving. This means a new deal in world economies, in trade between nations and the terms of that trade.

I believe that this can be done. But it is fundamental that there should be more wealth—more wealth to be shared more equitably. Shared more equitably, as you said, Mr Chairman in your welcoming address, within nations: but shared more equitably between nations and peoples.

How we fulfil that objective, by what measures or armoury of measures, we can decide. There are many means by which we can reach that end: but they are only mechanisms. Our dedication must be to the principle, and our determination to achieve it, is unshakeable.

My own concern and involvement with the problems of commodity trade is lifelong, but what is new is the extent of instability in food and raw material prices. Following the Korean War, which caused great disturbance, there was a long period of relative calm. But recently circumstances have been more difficult than at any time since the 1930s. Fluctuations in prices have been violent and sudden.

I would identify three reasons for these fluctuations:

First, variations in demand. In the last few years demand in the industrialised countries for the products of other countries has moved up and down more or less in step. The rise and fall has been greater than in the past and might well be even greater in the future. No-one, but no-one, has a vested interest in this sort of boom and bust. Commodity producing countries who gain from scarcity prices for a time, often find that inflation is generated in their countries, which causes great social hardship when the boom ends.

Second, variations in supply. There has been widespread disruption particularly from crop failures. The harvests of 1972 and 1974 were bad, as Mrs Gandhi reminded us yesterday. These bring disastrous consequences, above all for the poorest. Even now world food stocks are very low.

Third, of course, oil. The world economy as a whole suffered a sharp and unique jolt with the rise in oil prices.

These events underline one of the historic canons of one of the world's great religions, that "we are all members one of another".

Everything that has happened in these past two or three years demonstates the vested interest of all of us in a one-world system for commodity trade. We shall make no progress unless we recognise that large and sudden variations in price, not to mention uncertainty over supply and markets, are disadvantageous to both developed and developing countries alike. Both have a common interest in avoiding them.

The costs for the developed countries as consumers are a worsening of their inflation, setting up a ratchet mechanism of inflation as wages react up but not down: an extra burden on their balance of payments: and uncertainty over the long-term development of their sources of supply.

Britain's own balance of payments in 1974 had to carry an extra charge of £2,500 million by the rise in prices of oil, and commodity price increases over the past year represent a surcharge on every family in the country of 4 per cent of their household income. But developed countries do not gain all that much when commodity prices fall. They might hope for an improvement in their price level, but this does not always occur, particularly if the previous price inflation has generated a spiral of internal cost increases, wage and price increases. They might secure a temporary easement to their balance of payments, but it is only temporary, because one inevitable result is the impoverishment of their primary producing customers who have to cut down their imports of manufactured and other goods whether capital goods or consumer products. And this increases unemployment.

Before the War Sir William Beveridge and I produced evidence that every industrial slump in Britain, every increase in unemployment for the previous 100 years, was associated with a collapse in primary prices in the countries from whom we imported much food and materials.

And turning to developing countries, boom conditions can lead to excessive production and over-investment in capacity which may then prove uneconomic. In slump conditions while some developing countries suffer only from setbacks to their development plans, for many others it means malnutrition, even starvation. Less advanced countries depend critically on the ratio between their exports and imports of high-priced commodities.

And, again I repeat, those developing countries who neither produce their own energy needs nor raw materials for export suffer most whether in boom or slump. Especially those in absolute poverty.

What this analysis means, I can sum up in these terms.

We all have a common interest in reducing the violent fluctuations in commodity prices.

We must recognise the importance to developing countries of increasing their income from commodity exports.

The context for achieving these objectives must be an orderly and sustained expansion of world trade. Without this our joint efforts to bring order into trade in commodities will be frustrated.

Going back a little, the war released a new idealism and new intellectual resources. Bretton Woods, the work on the Havana Charter, which became permanent in the General Agreement on Tariffs and Trade, and the creation of the Food and Agriculture Organisation all brought forth a new sense of urgency. I was fortunate to have spent one of the most exciting periods of my life being involved with this. In 1946 my Prime Minister, Attlee, sent me as a young Minister to head the British Delegation to the FAO preparatory Commission. Sixteen nations were selected by FAO to prepare guidelines for the new organisation under two heads —primary production, particularly in developing countries, and commodity policy. Our report created a great deal of interest at that time—but little action. What we did do was to draft the guidelines for the new post-war international wheat agreement, for sugar and to seek to turn the pre-war producer cartels, tin and others, into genuine commodity agreements.

Under the then laissez-faire leadership of the United States, nations were content to make important decisions about tariffs, about freeing trade, and, through GATT, above all to lay down ground rules to prevent escapism by individual nations into autarky and into nationalistic measures harmful to their neighbours and to world trade in general. But the same philosophy proved allergic to vigorous action on commodities.

It was for that reason that my own Department, the Board of Trade, in 1951 took the lead in establishing the Commonwealth Sugar Agreement, a model of its kind, which has been of benefit to producer and consumer countries alike. One of my most savage criticisms of the terms on which Britain entered the European Common Market in 1971 was the acceptance of conditions which meant the ending of the Commonwealth Sugar Agreement. I am glad to feel that in Britain's renegotiations with the Community, we have at last got agreement to the guaranteed continuing import of the 1,400,000 tons of Commonwealth cane sugar into the Community, and it will go mainly to Britain. But for close on a quarter of a century on commodity markets, it has been "too much talk, and not enough do".

Each commodity poses a special problem. Each commodity has its own elasticity of demand, its own production cycle and its own special problems over storage. There is no general panacea. At the same time those who are charged with negotiating arrangements for trade in a particular commodity can assuredly benefit from adopting a common approach based on mutual undertakings and relevant mechanisms, some at least of which they might find appropriate to the issues they are seeking to resolve.

I want to say a word at this point—I am not running away from it—about proposals for indexation, for which many developing countries are pressing—namely the introduction of some form of indexation of commodity prices linked to the price of manufactured goods which developing countries import with the proceeds from their primary exports. This must certainly be looked at. It has been pressed by oil-producing countries, where technically its operation might be relatively simple, not least because there is basically one price for oil. Here the different starting points of producer and consumer countries are concerned with what date you choose—whether it is a peak oil price date, for example, whether you take the peak price or a figure so many percentage points behind it, or if you take the existing price whether you have a delay mechanism of a year or two or more before indexation begins to operate. In the case of general commodities the problems provide greater technical difficulties.

There is a problem of the starting date, the reference date. But not all commodities peak at the same time, or behave in a similar fashion following their respective peaks. Sugar producers, Mr Chairman, might like to take a date round about the latter months of last year when sugar was at its peak. I am not sure that President Kaunda would like to choose the same date in respect of copper, because by that date copper prices had fallen back to the 1970 levels, as though the commodity boom had never happened. So you have the problem of a common date for composite materials. I doubt whether it would be possible—and importing countries might not feel it desirable—to have an indexation based on a series of base points, each reflecting the peak points of individual commodities. And if you had, a sudden upsurge in one commodity price, or a catastrophic fall in another, would probably not be acceptable to the developing producer countries, since the benefit would be sharply differentiated, and might not be a general benefit at all. So if indexation is to be examined, as I concede it must, three considerations follow:

(1) We should take full account of the different kinds of indexation which are possible. Many countries, including oil producers, are talking about an index related to the cost of the goods they import—rather like a cost of living provision in a wages contract in our domestic economies. But in your own case, Mr Chairman, the methods you have adopted ensure that bauxite prices bear a fixed relation, not to import costs, but to the selling price of the finished aluminium.

(2) We have to recognise that indexation can bear very unequally as between the producer participants to a scheme, as well as to the consumer countries involved, and

(3) We must not let the technical difficulties of indexation, which may take considerable time to iron out, deflect us from the urgent necessity of examining the other proposals which I now want to put forward in my six-point plan, together with other proposals which might emerge from an examination of them.

Responding, Mr Chairman, to the challenge which you have placed before this Conference I now wish to place the British Government's views before you.

Any attempt to bring order to the key area of trade in commodities by following the approach I have outlined must build on the common interests of developed and developing countries. I believe that this common ground can be translated into a set of general commitments which would be complemented by specific proposals for action.

What the British Government have in mind is that we set as our objective a general agreement on commodities, not only for ourselves but for the whole world.

A generation after the General Agreement on Tariffs and Trade, I believe the time has come to balance it with a general agreement on commodities—it is long overdue.

As a basis for discussion, I suggest that the following commitments might form part of a general agreement.

First, we should recognise the interdependence of producers and consumers and the desirability of conducting trade in commodities in accordance with equitable arrangements worked out in agreement between producers and consumers.

Second, producer countries should undertake to maintain adequate and secure supplies to consumer countries.

Third, consumer countries for their part should undertake to improve access to markets for those items of primary production of interest to developing producers.

Fourth, the principle should be established that commodity prices should be equitable to consumers and remunerative to efficient producers and at a level which would encourage longer-term equilibrium between production and consumption.

Fifth, we should recognise in particular the need to expand the total production of essential foodstuffs.

Sixth, we should aim to encourage the efficient development production and marketing of commodities (both mineral and agricultural)—and I should like to emphasise forest products—and the diversification and efficient processing of commodities in developing countries. We should not deduce from two centuries of history that there was any divine ordinance at the creation of the world under which providence deposited the means to primary production in certain countries and it was ordained that those minerals, or other products should be exclusively or mainly processed in other countries.

In saying this I repeat that in any general agreement or other means to advance, we must lay heavy emphasis on the special needs of the poorest countries.

Now in practical terms, if we are to give specific content to these general commitments, specific action is called for. This action should, in my view must, include measures directed to the following ends:

(1) To establish better exchanges of information on forward supply and demand.

(2) To elaborate more specific rules to define the circumstances under which import and export restrictions may be applied to commodities.

(3) To encourage the development of producer/consumer associations for individual commodities.

(4) To give fresh impetus to the joint efforts of producers and consumers to conclude commodity agreements designed to facilitate the orderly conduct and development of trade. This could be done

First, by identifying commodities appropriate to the conclusion of such agreements;

Second, by analysing commodity by commodity the appropriate mechanisms for the regulation of trade within the framework of such agreements (including international buffer stocks, co-ordination of nationally held stocks, production controls and export quotas);

Third, by examining ways in which any financial burden arising from these mechanisms may be appropriately financed.

(5) To agree that the regulatory mechanisms incorporated in any international commodity agreement would be directed towards the maintenance of market prices within a range negotiated in accordance with the principles enshrined in the fourth general commitment.

(6) To establish the framework of a scheme for the stabilisation of export earnings from commodities.

These proposals will obviously need detailed study. There are already in prospect negotiations on a number of individual commodities including coffee, cocoa, tin and wheat. The United Kingdom has always belonged to previous commodity agreements and we shall play our part in the negotiations for these new agreements. Commodities will also be an important subject in the multilateral trade negotiations which are now getting under way and the European Community has stressed in its mandate for these negotiations its intention to take account of the interests and the problems of the developing countries and in particular of the least developed in all sectors of the negotiations.

Some commodities are of special if not exclusive importance to

Commonwealth producers, tea and jute, for example which have not shared in the recent commodity boom. Can we agree to tackle the problems of these commodities as a matter of urgency?

As my colleagues will see, in my specific proposals I have suggested that we look afresh at the possibilities for reducing price fluctuations. This could be done either through internationally held buffer stocks or through co-ordination of national buffer stocks, and you won't find that the answer that is right for one commodity is right for another. At the same time my proposals recognise that commodities produced by the poorest countries are not on the whole susceptible to price stabilisation agreements.

That is why I have suggested as a complement to price stabilisation, that we should examine schemes to stabilise export earnings. I propose this for a number of reasons. Such schemes are particularly helpful, one might almost say essential to countries where production is hit by drought or other natural disasters.

Twenty-two Commonwealth countries have had reason to welcome the Lomé Convention concluded between the European Community and the African, Caribbean and Pacific (ACP) countries. These negotiations were not initiated as a result of the British Government's renegotiations, but they were powerfully accelerated and intensified by the effect of our renegotiations. I have made clear to our own Parliament that we intend to build on Lomé in respect of Asian countries, so powerfully represented at this Conference—Asian countries who have gained a little but have not participated in the arrangements for ACP countries.

Under the Lomé Convention forty-six countries will begin to benefit from the export stabilisation scheme, STABEX. This is an important step forward, though limited in scope. It may be difficult to extend it in its present form, and of course International Monetary Fund (IMF) schemes are also limited in scope. That is why we have to look at other possibilities for wider self-financing schemes.

Mr Chairman, you called on Tuesday for a new approach, for the reversal of generations of world history, in which the principles so many of us have advocated for greater equality within our countries could become a reality as between countries, a reality which would put an end to the centuries old exploitation of so many primary producers by so many importing manufacturing countries. That was what you charged us with on Tuesday morning.

These principles I accept. Indeed I accepted them when, together with others, I founded that movement in Britain, whose work has become known in the Commonwealth and more widely, "War on Want".

In your inaugural speech on Tuesday, and equally in your interview on British television on these matters which so greatly impressed many of my fellow countrymen, it was clear that you were at that time asserting the arrival of a new age in world economic affairs. The producing countries were going to strengthen and exercise their muscle, indeed some were already doing so, and they were going to do that as a counter to the muscle which you and others feel has been exerted against primary producing countries by those who take their products—and who in the past have themselves not disdained the use of cartels, both in buying and selling. But in that television broadcast you expressed a clear preference that once a balance of power had been established, then the solution should be sought through reason and argument and the creation of advantageous agreements covering all aspects of commodity stabilisation.

The same thought clearly inspired an important passage of your speech on Tuesday:—

"The choices that face us are these: producer associations can either become the instruments through which producing nations conduct a rational dialogue with consumers within the framework of a new economic order: or, for want of dialogue, they will become increasingly the instruments through which the third world takes such unilateral action as is demanded by the cause of survival and equity. The choice, therefore, is inevitably between dialogue and confrontation. The challenge to this Conference is to explore the ways by which the scales of probability may be tipped in favour of dialogue".

And it was in this context that you used the phrase which moved us all, when you said, "I believe that the Commonwealth may be uniquely blessed for this effort".

For this Conference represents over a quarter of the membership of the United Nations and a quarter of the world's population: it represents every continent, every ocean, every significant region of the world. It should neither have come here nor should it leave with an inferiority complex about its moral power, and its power to give a lead in world affairs.

Naturally, we cannot negotiate a general agreement on commodities here at Kingston. All I am hoping for is a "Second Reading Debate" on these proposals. If my colleagues saw merit in the general approach, perhaps Commonwealth countries could together carry the Debate forward at the 7th Special Session of the UN General Assembly in September next as a contribution to the Debate on world economic development, so that practical work could be set in hand when the 4th United Nations Committee on Trade and Development (UNCTAD) meets in another Commonwealth capital, Nairobi, next May.

But in what you have proposed, and in what I have attempted to set out, there is a greater degree of common ground than many would expect, recognising that we speak for countries with a very different economic background—come here to link hands across the so-called great divide between producers and consumers, advanced and developing countries.

In a purely domestic speech in Britain last year to our trade union movement, dealing with Britain's internal problems, I said, referring to the forthcoming American Bicentenary celebrations, that what we needed in Britain, as in the world, was not a declaration of independence, but a declaration of interdependence.

Your theme, supported by others this week, was that the great battle for political independence of previously dependent nations has been largely won and that the struggle now is for economic independence for those nations and peoples. But what we are both saying is that independence is not enough, indeed that economic independence can turn out perhaps to be itself no more than a fantasy. What we both are seeking, and I believe this Conference, are seeking, is mutual advantage and mutual concern, based on each seeking our individual strength, with a framework of world interdependence.

MEMORANDUM

WORLD ECONOMIC INTERDEPENDENCE AND TRADE IN COM-
MODITIES

The Purpose of this Memorandum

1. One of the world's most pressing and complex problems, which
   faces all governments, is trade in commodities. The survey of commodity problems in this paper is intended to assist the Conference
   in their discussion of this issue.

2. This first section of the survey begins with a review of commodity
   problems in the interdependent world of today. This is followed by
   a brief account of international discussions which have recently
   been taking place to try to provide a framework within which the
   problems of commodity trade can be sensibly tackled. The survey
   then goes on to describe the history and operation of commodity
   agreements, and to assess the scope for future co-operation in the
   stabilisation of prices and supplies, including schemes for the
   stabilisation of earnings from commodity exports. Some of these
   issues and several related questions are dealt with in greater detail
   in the supporting material in Part Three of this White Paper.

Commodity Problems in a World Context

3. The need for co-operation by all members of the international community is clearer now than it has ever been. World-wide inflation,
   coupled with deepening recession in many major countries, massive
   imbalances in international payments and rapid movements in
   exchange rates have combined to pose problems for all governments.
   The interdependence of nations is such that no government can
   find solutions to these problems on its own.

4. The events of the past few years have highlighted the special difficulties associated with internationally traded commodities: food and raw materials. In no field of economic activity is the interdependence of the world community more obvious than in the production, trade and use of commodities. All countries, even those most richly endowed with natural resources, rely to some extent on imports from other countries for an adequate supply of many of the commodities they need. All countries, even those with the fewest natural resources, produce some commodities. All countries are concerned in commodity trade and many are vitally dependent upon it. It is impossible to make simple distinctions, such as between poor producers and rich consumers. The problems of commodity trade can only be solved by co-operative international action.

5. This overlap between producers and consumers means that all countries are likely to be affected by instability in commodity prices, though in different ways and to varying extents. Many countries' balance of payments were disturbed by the recent commodities price boom; the five-fold increase in the price of oil seriously affected all non-oil producing countries; the effects of each were accentuated by the combination of the two.

6. Many developing countries depend largely on the ratio between their earnings from commodity exports and their expenditure on commodity imports. Those with a strong export balance in commodities are better able to withstand commodity increases, and even the oil surcharge. But they may still lose on imports much of what they gain from increased export prices. Countries with few or no commodity exports will inevitably be hard hit by increased prices. Some developed countries, including the United Kingdom (until North Sea oil is fully on stream), fall into this category. Still harder hit by price increases on their imports of energy and of essential foods and raw materials are the developing countries with no exportable commodities; while those countries which were already facing starvation from famine or drought or for other reasons have been most seriously affected of all.

7. Instability in commodity prices, gluts and shortages, are not in the long-term interests of developing international trade. They make long-term planning of production difficult if not impossible. Neither producers nor consumers are likely to gain from violent market fluctuations. Producer countries which may seem to benefit for a time from very high prices often find that they too are subject to the inflation that is generated. Moreover, demand may be reduced and substitution encouraged. Similarly, a serious slump in prices is not in the long-term interest of the consumers. Production is likely to be curtailed and scarcity induced. Industrialised countries which import commodities but rely on the export of manufactures may

suffer as much from the impoverishment of their overseas customers among the primary producers as they benefit from lower
import prices. In the long term, both producers and consumers
share an interest in orderly trade in an expanding world economy.

8. Recent experience has shown how far we are away from orderly
and stable trade in commodities. Even if we leave aside the special
case of oil, the simultaneous upsurge in demand in 1972/73 in most
industrialised countries was followed by massive increases in the
price of food and raw materials which gave renewed impetus to
world inflation and dislocated balances of payments with effects
that are still only too evident. For some commodities the subse-
quent price fall has been such that the expansion of supplies to
meet future world demands is jeopardised. If we neglect the lessons
of the past, we may face higher peaks and deeper troughs in the
future.

9. In part, the lessons have not been learned because the problems are
so complex. Each commodity is a special case, with special features
affecting its production and use. Some have a long production
cycle, some are perishable, some are bulky, some are in competition
with synthetics, and so on. Indeed, although we can detect general
commodity cycles in the statistics, there are major variations in the
experience of individual commodities. Even now, the adjustments
to the boom are still working their way through. It is important to
recognise this diversity, so that we are not misled into thinking
that there is any easy solution of general applicability. But that
diversity is also a challenge to our ingenuity and our will to co-
operate in working towards solutions.

The Scope of International Discussions

10. There is no straightforward equation between producers and developing countries on the one hand and consumers and developed
countries on the other. Nevertheless, it is true that trade in primary
products is of greater relative importance to developing countries
and it is therefore understandable that over the years, and particularly in the last few years, the developing countries should have
devoted special attention to the problems of trade in commodities.
In a series of international meetings and at meetings of their own,
developing countries have set out in some detail their views and
proposals in this field as in others. Their proposals covered a wide
range including commodity agreements; buffer stocks; equitable
prices for exports and safeguards for export earnings; indexation of
commodity prices; improved access and conditions for developing
country products in the markets of developed countries; commo-

dity producer associations; and help for expanding the capacity of developing countries for processing commodities for export.

11. Some of these proposals have not been acceptable to developed countries and inevitably there have been disagreements which have prevented progress towards the practical application of new arrangements for commodities. A central point of difference hitherto has been that in general the developed countries have considered that work should proceed on a commodity by commodity basis while the developing countries have usually formulated requests for wide-ranging arrangements covering all or many commodities.

12. These differences of approach should not, however, be allowed to obscure the fact that there is a great deal of common ground and and common interest. The events of the last two or three years have led to new thinking and a readiness to re-examine long-standing problems with fresh minds. The recent meeting of the UNCTAD Committee on Commodities gives encouraging evidence of how common ground can be built upon. The chief interest was in the UNCTAD Secretary General's proposals for a programme of work on the "integrated approach" for commodity problems. The resolution adopted by consensus calls for further studies of this. The "integrated approach" is based essentially on the idea of setting up an internationally-financed stockpile for 18 diverse commodities, (Wheat, Maize, Rice, Sugar, Coffee, Cocoa, Tea, Cotton, Jute, Wool, Hard fibres, Rubber, Copper, Lead, Zinc, Tin, Bauxite/Alumina, Iron-ore), linked to a system of multilateral commitments for sales and purchases together with compensatory financing and an increase in local processing. A great deal of work has been done both by individual countries and in international bodies to study all these problems in depth. The study by the Commonwealth Secretariat on Terms of Trade Policy for Primary Commodities is a most useful contribution. Both the new climate of thought and the detailed work which has already been done provide a new and encouraging basis on which to build. We must still bear in mind that each commodity has its own special characteristics and that any multi-commodity arrangement would be difficult to devise. However, it may well be sensible to search out and identify a number of common features which apply to most commodities and to see whether these can be brought together to provide a new and more comprehensive approach to the whole problem of commodity trade.

Commodity Agreements

13. Since well before the Second World War attempts have been made to regulate trade in commodities by means of mechanisms aimed

at stabilising supplies and prices. So far consumers and producers have succeeded in reaching agreements with economic provisions to cover only a handful of commodities. The evidence suggests that these agreements have been more successful in protecting floor prices than ceiling prices. Moreover, they have proved incapable of dealing with the kinds of fluctuations in price and supply that have occurred over the past few years.

14. Despite this, we should be prepared in suitable cases to seek to negotiate new agreements, and to try in these to avoid the short-comings experienced in the past. The first step should be to identi-fy which commodities represent the most likely candidates for new or renewed arrangements. Here the work of the FAO the GATT and UNCTAD is of considerable value. Our initial assessment of those commodities for which new or renewed arrangements would be most likely to help in assuring adequate supplies at reasonable prices and in improving the conditions of the poorest countries, suggests the following possibilities: cereals, cocoa, coffee, copper, dairy products, jute, rubber, sisal, sugar, tea and tin. Of these tea, jute and sisal are of particular interest to the poorest Common-wealth producers.

15. In negotiating each commodity agreement it will be necessary to re-examine the whole spectrum of mechanisms for stabilising price and supply. This is well-trodden ground and no radically new strategy on price adjustment mechanisms is likely to emerge; terms have to be negotiated for each commodity and the constraints are well known to consumers and producers alike. The issues are con-sidered in more detail in Part Three. The stabilisation of prices and supplies has traditionally been attempted by means of export/im-port quotas and buffer stocks. Major problems here are the high initial cost of establishing even a modest level of stockpiling and the effect of high interest rates on the cost of maintaining stocks. The developing countries have already put emphasis on the need for financial help to establish stockpiles; and in this area there could be mutually beneficial arrangements between producers and con-sumers. The difficulties involved tend to be specific to individual commodities and these are outlined in the appropriate annexes, but there are also general problems such as the exercise of control over the management of the stocks.

16. While recognising the need to try to maintain and where possible to increase the real income of developing countries from their primary exports, there are very great problems involved in attempting to establish any automatic link between the price of particular com-modities and the price of manufactured goods. Such a system would be inflexible and could not be guaranteed to produce increased earnings for the producers without the most elaborate arrangements

between the governments of all the main producers and consumers. There are the obvious risks of substitution if prices are raised above the levels which the market will bear, as well as those of reduced consumption. If such a system were to be fully effective it could reinforce inflationary pressures in times of high demand and add to the difficulties of developing countries through the effect on their own import bills.

Export Earnings Stabilisation

17. Commodity agreements offer the possibility of bringing about greater stability in commodity prices. The experience of agreements in the past suggests that over a period they will tend to maintain prices at somewhat higher levels than would otherwise obtain and to that extent to bring about some degree of transfer of resources from consumers to producers. But not all the producers would be developing countries and some of the commodities produced by the developing countries may be particularly difficult candidates for effective commodity agreements. A complementary approach would be to extend the scope of schemes for stabilising the export earnings of developing country producers. Among the advantages of earnings stabilisation schemes are:

   (a) as they do not attempt to operate directly on price levels, they are less affected by fluctuations of supply and demand than are prices stabilisation agreements;

   (b) they can be specifically related to the problems of individual countries;

   (c) they can benefit individual producers suffering from temporary disruptions in supplies.

18. Several schemes for increasing the stability of export earnings amongst producer countries have been put forward in the past. At the present time there is one in operation under the aegis of the IMF and another, resulting from the negotiations between the European Economic Community (EEC) and ACP countries in the Lomé Convention, will take effect from the entry into force of that Convention. Although there is clearly no fundamental difficulty in devising broader schemes, there are limitations on their effectiveness.

19. In looking at this complex of issues we must continually have in mind the special problems of the poorest developing countries which have been hard-hit by the impact of the oil crisis, which have themselves suffered from the increases in the price of fertilisers and of food and other raw materials, which have not necessarily shared

in the boom in raw material prices, and which now face the threat of reduced demand for their products in a world recession. They need adequate, reasonably-priced and increasing supplies of feeding stuffs and of fertilisers to produce more food, to enable them to raise themselves above subsistence level and to develop exportable products. Many of their problems must be tackled through conventional aid. But it may also be necessary to devise favourable terms for them in any scheme of export earnings stabilisation or in relation to the financing of buffer stocks of commodities of particular interest to them.

Conclusion

20. There is no neat and tidy solution to the commodities problem which will satisfy all the aspirations of developing countries on the one hand and be wholly acceptable to the developed countries on the other. International discussion over a period of many years has given rise to a bewildering variety of proposals. We must create an atmosphere of greater confidence, in which we can all look forward to a more assured future. A first task might therefore be to draw up an agreed statement of the common ground between consumers and producers, whether developed or developing, whether rich or poor.

21. Such a statement might take the form of a general agreement on commodities, containing general commitments and specific proposals for action, as set out below—

General Commitments

1. To recognise the interdependence of producers and consumers and the desirability of conducting trade in commodities in accordance with equitable arrangements worked out in agreement between producers and consumers.

2. On the part of producer countries, to maintain adequate and secure supplies to consumer countries.

3. On the part of consumer countries, to improve access to markets for those items of primary production of interest to developing producers.

4. To establish the principle that commodity prices should be equitable to consumers and remunerative to efficient producers and at a level which will encourage longer-term equilibrium between production and consumption.

5. To recognise in particular the need to expand the total production of essential foodstuffs.

6. To encourage the efficient development, production and marketing of commodities (mineral or agricultural, with emphasis on forest products) and diversification and the efficient processing of commodities in developing countries.

In any general agreement, or other means to advance, heavy emphasis must be laid on the special needs of the poorest countries.

Proposals for Action

1. To establish better exchanges of information on forward supply and demand.

2. To elaborate more specific rules to define the circumstances in which import and export restrictions may be applied to commodities.

3. To encourage the development of producer/consumer associations for individual commodities.

4. To give fresh impetus to the joint efforts of producers and consumers to conclude commodity agreements designed to facilitate the orderly conduct and development of trade. This could be done, first by identifying commodities appropriate to the conclusion of such agreements; second by analysing commodity by commodity the appropriate mechanisms for the regulation of trade within the framework of such agreements (including international buffer stocks, co-ordination of nationally held stocks, production controls and export quotas); third by examining ways in which any financial burden arising from these mechanisms may be appropriately financed.

5. To agree that the regulatory mechanims incorporated in any international commodity agreement would be directed towards the maintenance of market prices within a range negotiated in accordance with the principles enshrined in the fourth general commitment.

6. To establish the framework of a scheme for the stabilisation of export earnings from commodities.

ADDRESS AT THE COMMONWEALTH HEADS OF GOVERNMENT MEETING 1975, ON THE NEW INTERNATIONAL ECONOMIC ORDER

Forbes Burnham

I

The dialogue on the international economic situation is essentially a dialogue between the developed and the developing world, between the rich countries and the poor, between the few and the many, between the strong and the weak, between those at the centre of economic power and those on the periphery.

I speak, Mr Chairman, from the position of the developing world without apology and, as I shall try to do, with frankness and candour as among friends.

In a more specific sense, I speak not only on behalf of Guyana but of others in the Caribbean. In part, therefore, Mr Chairman, on your own behalf.

As Member States of the Group of 77, we naturally approach this discussion with the perspectives of the Third World and all that I shall say will be conditioned, by the joint efforts in which that Group has long been engaged, and by the aspirations which are common to us all.

To these efforts and aspirations, all of us around this table who are developing countries, have consistently contributed and subscribed. In a very real sense they found expression in the decisions of the Sixth Special Session of the General Assembly almost exactly one year ago. They are the essence of the commitment which we bring to this discussion.

Against this background, and from these perspectives, what is the global economic situation of which we speak?

It is one in which the polarization of wealth and poverty between countries, and the economic cleavage between the north and south have steadily increased.

This alone would be sufficient to merit the description of global crisis, but the situation is further compounded by the fact that such sources of stability as once existed have been swept away.

The International Development Strategy had been founded on the assumption that national processes for development would, for the most part, continue uninterrupted and that what was necessary was mainly to commit the developed societies to a significant increase in their aid contributions.

Now, in the middle of the Second Development Decade, as the International Development Strategy nears its mid-term review, that Strategy is of doubtful relevance and seems to be an exercise in futility.

The International Order within which it was conceived is rapidly disintegrating. Today, the whole system is in disarray. Some of its institutions are of questionable significance and value and must clearly be rebuilt.

II

The Lack of Political Will

While, in some quarters, there is a new consciousness of the limits of resources and shortages especially in the area of food, there is lacking nearly everywhere in the developed world the political will to translate such awareness into policies which would share the world's wealth more fairly.

Despite the temporary setbacks of the energy crisis, the developed world has continued to pursue its affluent way of living while on other parts of this planet it is estimated that eighty thousand innocent children die of hunger each day.

A distinguished Third World Economist Mr Mabbubul Haq of Pakistan, now Director of the Policy Planning and Programme Review Department of the World Bank, has recently assembled some of the statistics which illustrate the misery of the greater part of the human race:

> "[It is] a world . . . so divided economically as to have about 20% of the population enjoying about 80% of the world income.
>
> We have today about two-thirds of humanity living — if it can be called living at all — on less than 30 U.S. cents a day.
>
> We have today a situation where there are about 1,000 million illiterate people around the world, although the world has the means and the technology to spread education.
>
> We have about 60% to 70% of the children in the Third World suffering from mal-nutrition, although the world has the resources to give adequate nutrition to all of its population.
>
> We have mal-distribution of the world's resources on a scale where the developed countries are consuming about 25 times more of the resources per capita than the developing countries.

We are in a situation where in the Third World millions of people work incredibly hard for very miserable rewards.

It is easy to be very sophisticated about it and to explain it all in terms of stages and development but it is not likely to carry much conviction in our countries where people toil in a broiling sun from morning till dusk for mere subsistence and for premature death without ever discovering the reason why."

While the disorder grows and poverty widens, in several international fora detailed blue-prints of an integrated and just world order have been worked out.

III

The Possibilities of Confrontation

Approximately a year ago the Sixth Special Session of the General Assembly adopted a comprehensive Declaration and a Programme of Action which provided a blue-print for a new International Economic Order.

The documents were projected to the world as representing the consensus of the international community. But while the programme had overwhelming support, we must frankly recognise that the consensus had been eroded at the beginning by the important reservations of several developed countries and the lukewarm support of others.

Some, hardly veiled their almost emotional opposition of that sovereign act of development by developing countries of nationalising, owning and controlling their natural resources.

Since then, although there has been increasingly wide recognition of the need for fundamental change in the relations between the developed and developing countries in the key institutions of the system, there has been little attempt to search for areas of agreement and to implement a new programme.

Instead, all the signs point to ominous possibilities of confrontation. Indeed, the trends are disturbing.

Several major countries and groups of countries are pursuing, as if unaware of the trends of change, policies which may earn them short-term benefits but can only lead to global catastrophe.

In Dakar, in February this year a hundred developing countries, a large number of them the producers of raw materials, met to consider joint action and to agree on a basis which would ensure that producers would be effectively represented at the meeting proposed by the French Government between industrial consumers and the OPEC countries.

Little, if any, account appears to have been taken of the agreements reached and the resolutions passed at Dakar.

In the event, however, the preparatory meeting ended in deadlock arising mainly from the refusal of the developed States there represented to contemplate a dialogue beyond their immediate pre-occupation with oil.

It is only in one area that an important advance has been made. This is the new style relationship which has been negotiated by the forty-six developing countries of Africa, the Caribbean and the Pacific with the European Economic Community.

One is aware that the Lome Convention is not perfect and is only a beginning — an outline which must be filled in and refined by the negotiation of further areas of agreements and by practice.

There are important observations to be made in the light of this development:

(a)   first, while it clearly demonstrates the possibility of evolving new and in some degree just economic cooperation arrangements between developed and developing countries, we must ensure that such group gains do not become a factor for division, but are globalised as soon as possible for the benefit of all developing countries and the international community as a whole;

b)   second, any significant new relationship must not be limited only to trade in one or more commodities but should be multi-dimensional, including importantly the creation of new economic opportunities through, for example, the development of economic cooperation among the developing countries themselves, as well as the ancilliary instrument of aid;

(c)   third, and immediately relevant for us, Commonwealth links contributed to both the genesis and the basis of solidarity on which the ACP Group was built. Indeed, the Group owed its origin to an initial informal meeting of certain Commonwealth Foreign Ministers held in Georgetown, Guyana in August, 1972 at the time of a Conference of Foreign Ministers of the Non-aligned Movement.

IV

Commonwealth Cooperation

As an association of countries drawn from every region of the world, representative of several levels of development and strategies of development, composed of peoples of every ethnic group and of a diversity of cultures and creeds, the Commonwealth is, so to speak, a "sample" or microcosm, as some would put it, of the international community.

There is the further advantage that there is no super-power within our group.

In our intimate consultations, we are less subject to public pressures.

It can, therefore, be our Commonwealth vocation to advance Commonwealth cooperation in ways which render service to the international community as a whole in its search for areas of agreement and for concrete steps and measures through which the new International Economic Order may take shape.

V

The Need for a New Order

It will not do to deceive ourselves about the nature of the task which must be attempted. **We must embark on building a new order. It is not a task involving the repair, renovation or piece-meal reconstruction of the old order.**

The sharp rise in petroleum prices did not precipitate the crisis of change. It illuminated its dimensions and its inter-related aspects.

So, let us begin by exploring the considerations which must inform the process of implementing a new order:

1. first, global considerations must prevail over narrow national interests, not only because of moral convictions about the brotherhood of all mankind, but, also, because of the perception that planetary resources are limited.

   If the volume of resources cannot be increased indefinitely, this means that existing resources must be so divided that the poor get a larger share of them. Or, to put this another way, the problem is not only one of ensuring stability of markets, adequacy of supplies and fair prices, but one of rational use and resource management.

   One of ensuring that the rate of consumption of resources is related to the needs of the people who own the resources and of mankind as a whole, and that resources are utilised in an optimum way.

   Thus, many of our oil producing friends may be right in contending that in the foreseeable future the main use of oil should be in the production, not of energy, but of petro-chemicals.

2. second, we must transfer and implement, at the international level, the insight which we already recognise at the national level — namely, that growth alone will not provide automatically the jobs and opportunities for human development which our people demand and deserve.

   The "trickle down" approach must be abandoned at the global level.

   International development and national development must, therefore, be made purposeful.

   This must be done through deliberate planning and special mechanisms which ensure that the gains of the global system are equitably distributed at all levels.

VI

The Perspectives for Change

At the same time, we must recognise that the old order was, and still is, supported by a concentration of economic and military power in the developed world; was, and still is, exercised through an infra-structure of institutions and of transport and communication links.

This international economic system assigned a peripheral role to the developing countries. Therefore, a move to a new economic system which is equitable must at the same time seek to bring about a redistribution of economic strength.

It is only the mobilisation of strength by the OPEC countries which has led to the serious consideration by the developed countries of the possibilities of change.

It is only the continuing use of such countervailing pressure which will ensure that the need for fundamental change remains on the international agenda.

In the redistribution of economic power, we are convinced that the chief instrument should be horizontal action among the developing countries themselves, not only within regions and sub-regions but inter-regionally.

Such action should include joint action among producers, the development of trade and transport and communication links, joint industrial enterprises and the transfer of technology.

In this connection, the Secretary-General, pursuing the suggestion of the late Norman Kirk, has drawn attention in his report to possible economic complementaries of the Commonwealth countries which look out on the Indian Ocean and the Pacific. (*See page 19, para. 53 of the Secretary-General's Report*).

There are also obvious possibilities of other trade linkages such as between Africa and the Caribbean.

VII

A Comprehensive Action Programme

There is another danger that needs to be guarded against if we are all serious in our commitment to programmes of positive action which will give reality to a new international economic order.

It is the danger of deceiving ourselves that we can somehow achieve fundamental change by marginal adjustments and devices of a piece-meal and reformist nature.

This is not to say that there is no value in one particular approach or another. It is to emphasise that we shall not make real progress unless we evolve an integrated programme designed to fulfil, not merely the aspirations of the developing world, but, also, the necessities for survival of the global community.

In such a programme, there may be a room for particular approaches, but their value would lie in their being part of a workable programme which, because of the linkages within it, offers a chance of lasting results.

The Secretary-General of UNCTAD has himself emphasised the essentiality of such an approach in which, for example, buffer stocks, commodity arrangements, compensatory financing schemes for the stabilisation of export earnings, indexation, producers associations – can all have their place.

If the Commonwealth is to make a real contribution to advancing the international dialogue on the implementation of a new international economic, order it is to such a comprehensive programme of action that its most eminent minds must be directed.

We believe that we have within the Commonwealth the human resources that can be so deployed. What is needed from us here at Kingston is the political will to direct that they should be engaged upon this task.

Furthermore, a New International Economic Order which has as its objective just and equitable economic relations in the international community does not begin or end with commodities.

Our concern is not only about the market place. The New International Economic Order must be predicated upon an international economic environment which is capable of accommodating different economic and social systems.

Moroever, we should not talk about commodities without talking about the transfer of real resources, without talking about the transnational corporation, that instrument of exploitation, and the transfer of technology, and without recognising the urgent need for the reform, nay restructuring, of the appropriate international institutions.

We must remark the critical role that must be played by the international monetary and financial system on which I hope others in this debate will enlarge.

VIII

The Option of Cooperation

Mine is the attempt to emphasise that we cannot allow ourselves to be diverted from the main thrust towards a new international economic order involving an integrated programme of change.

The urgent need is for the Commonwealth to make its contribution to this task by entrusting to its best minds an examination of the possibilities of achieving such an objective.

It is to this end, therefore, that I wish later to make a proposal for endorsement hopefully by my colleagues.

Of course, it is not enough to identify the parameters within which the New International Economic Order must be built.

The other fundamental question before us is whether it will be brought to birth through confrontation and conflict, with their attendant difficulties for the peoples of both the developed and developing countries, or whether the new order will be built through consultation and consensus.

Fortunately, there are some encouraging signs. In Europe, both inside and outside of the Community, a few countries, and in particular the Netherlands and Sweden, have demonstrated their commitment to the New International Economic Order.

In addition to their support for national programmes dedicated to structural change, these two countries have provided enlightened support for an interregional Project for which the Guyana Government is responsible to the Non-aligned Movement and which has as its objective the development of trade, joint industrial enterprises and transport links among Non-aligned and other interested developing countries.

In the Commonwealth, given a recognition of the realities and the need for the mobilisation of political will, we can do much to ensure that the second way of consultation and consensus prevails.

We have been particularly encouraged by the assertion by Canada of the need for a global ethic and of the urgent need to create a just order.

On the other side of the globe, the commitment of the Governments of Australia and New Zealand to Third World causes can likewise provide us with a basis on which a Commonwealth consensus can grow.

Here, in this milieu of frank and easy consultation, there is no need for the rhetoric which can so often lull us into the mistaken belief that action is already in train. Nor will assurances be sufficient.

We are long past the stage when such would suffice, unless, of course, such assurances are translated into guarantees and commitments to structural change which can contribute to development in a practical way.

IX

A Multi-selective Approach

It will be clear, Mr Chairman, from what I have said, that developing States are bound to question the validity of an approach to the installation of the New International Economic Order, for which the Third World, and some at least of the developed countries, are working, through the selective mechanism of a special agreement on commodities alone.

I have listened with great interest to the proposals put forward by our good friend and colleague, Harold Wilson.

As he has rightly said "we cannot negotiate a general agreement on commodities here at Kingston". Of course, we cannot — because that is a matter for the international community as a whole.

Nor can we pre-empt the debate upon it by premature endorsement, or conversley, by premature rejection.

What I can say for our part, is that we cannot but look askance at piecemeal proposals which confine themselves to merely one area of the Programme of Action by which the New International Economic Order must find fulfilment.

In this respect, it is obvious — and I am sure Mr Wilson will agree — that his proposals do not, and indeed do not pretend to, constitute a comprehensive response to the Programme of Action for which the General Assembly has called.

As the Secretary-General of UNCTAD has emphasised, that Programme will remain frustrated unless approached through an integrated series of mechanisms.

And, as has been demonstrated in a different context, it is seldom that a step by step approach is either right or efficacious when seeking solutions to truly fundamental problems.

We must, and I am sure I speak for all of us, thank Harold for the care with which he has advanced his proposals: but I owe it to him and to the Meeting, to indicate our initial anxieties about them.

X

British Proposals and the Lessons of Yesterday

Let me express, immediately, my concern that the proposals which Harold has made with respect to commodity arrangements do not match the objectives which his own Government and he have set themselves.

He has explained to us that the guiding principle underlying his proposals is "that the wealth of the world must be re-distributed in favour of the poverty-stricken and the starving", and he went on, Mr Chairman, to quote from your inaugural speech in voicing his own support for new measures which will bring about such a re-distribution.

I was, therefore, rather puzzled to find that his proposals seemed to envisage merely mechanisms for the stabilisation of prices and earnings rather than for substantially increasing real earnings in the interest of re-distribution.

In that sense, the scheme corresponds more closely to the concepts which underlay stabilisation arrangements of the 1950s rather than the new perspectives that are being entertained today.

The system of the '50's was constructed in 1945 immediately after the Second World War. In 1945, many of us were still at school, in 1975 our grey hairs are going down to the grave in poverty and dissillusion-ment.

In 1945, the United Nations had 41 mostly rich members. In 1975 it has 137 mostly poor members; yet we are asked to accept and assume the relevancy of the concepts and institutions put forward and created respectively by the few rich thirty years ago.

We are handed on sacred tablets, for example, rules and principles of international law laid down by the rich to protect and preserve their interests and predominance.

How can this serve the objective "that the wealth of the world . . . . be redistributed in favour of the poverty-stricken and the starving"?

XI

The Contemporary Realities

If we are, in the contemporary world, to settle guiding principles which would govern our efforts to narrow the gap between the incomes of the primary producing and the industrialised countries, then we must modify Harold's six principles in at least four respects.

First, we must agree that the terms of trade between primary producers and manufacturers would be the product of the conscious planning of relative movements in real income as compared to mere reliance upon the free inter-play of market forces.

Second, in our thinking about specific arrangements for primary pro-ducts, we must be careful, in referring to producer-associations, to avoid pejorative terms like "cartels".

These are essentially exploitative institutions native to developed in-dustrial countries and have no propinquity with the producer association designed to save the poor from being further impoverished.

The fact is that, if primary producers have to face selling situations where buyers continue to be in a position to exercise a considerable amount of market power, then there is no alternative but for the former to seek to increase their market power by forming producers associations.

Third, as an extension of the first and second points, we must be pre-pared to re-examine concepts which have formed a part of traditional thinking on commodity trade.

For example, it is by no means clear that old understandings concern-ing the meaning of terms such as 'prices being equitable to consumers' and 'remunerative to efficient producers', are appropriate in the contem-porary situation where the principal interest to be safeguarded is that of the national economy rather than that of the private consumer or pro-ducer.

We must be careful not to treat as moral absolutes, concepts developed in the context of a particular economic and social system and in the context of a specific historical experience.

Fourth, it can be readily conceded that in our thinking about commodities, we must give recognition to consuming interests.

However, we may be deflected from our main objective if we seek to introduce, as a condition of our new efforts in this field, formal and built-in guarantees concerning the security of supply to consuming countries.

In the end, this security can be best achieved through satisfactory trading arrangements based upon some of the precepts I have advanced earlier.

XII

A System of Indexation

Turning to the question of mechanisms for indexation, there should be none among us who does not recognise the technical complexities of the question.

However, some of the difficulties referred to by my colleague seem to me more apparent than real.

In the first place, it is difficult to understand his specific concern about the matter of reference dates. There is no fundamental reason why there cannot be different dates on the basis of which reference prices can be calculated; and, in any case, a reference price need not be a historical price.

Moroever, if I may speak on behalf of the bauxite producers, the new method of valuing bauxite was mainly designed for the purpose of finding a more acceptable basis for the taxation of the ore than is provided by the arbitrary transfer prices fixed by the transnational bauxite companies.

It is, therefore, not strictly correct to describe the bauxite arrangements as a scheme of indexation.

But to return to indexation — contemporary experience in the field of international trade has revealed all too clearly the need that has always existed for commodity arrangements to be put on a just and equitable basis through a system of indexation in respect of commodity prices.

A system — whatever the actual mechanics — that would ensure for producing countries prices that were just in a comprehensive sense — just in relationship to the prices which these countries must themselves pay for the manufactured goods they need to import; just in terms of moving toward an egalitarian world society in which the living standards of the workers who till the fields and shovel the ore in the producer countries do not remain behind the living standards of those who work in the factories of the industrialised States.

Unless this need is met, we remain in danger of seeking solutions through mechanisms which, at the optimum, do no more than preserve existing disparties.

Regrettably, but perhaps not surprisingly, the concept of indexation has found little favour in the developed world despite the fact that there has been a long history within those States of the use of mechanisms of indexation for implementing policies of equalisation within their own societies, notably between the farming and non-farming sectors.

What is clearly needed is for the best minds to be applied to translating the objective of indexation into workable mechanisms.

It is to this task that we should pay the most urgent attention and Commonwealth action could contribute in no small measure.

XIII

Specific Proposals and the Realities of Tomorrow

This is not the place for me to respond in detail to other points in Harold's presentation which I feel warrant qualification and even re-for-mulation.

However, as a general observation, I think that, as we attempt to devise proposals for solving problems in this field, we ought to exercise the greatest care in ensuring that there will not be differential in their impact among developing countries — notwithstanding the special arrangements which must be made for categories such as the least develo-ped.

Nothing can be more damaging to the efforts at co-operation between Third World countries which all of us in this room are anxious to promote.

If we accept the thesis that we would render a dis-service to the evol-ving international dialogue by appearing here at Kingston to settle, how-ever preliminary or in principle, any arrangements that leave entirely untouched innumerable areas of the total Programme of Action to which so many of us are committed, how then shall we proceed?

I suggest that progress can best be made, in Commonwealth delibera-tions as in the international dialogue, only through a comprehensive and integrated approach.

If the Commonwealth has a role to play in contributing positively to the search for solutions, that role must proceed along such lines. It is a role that must respect this principle.

We believe that there is need for such an effort which I suggest might best be undertaken on the basis of the following approach —

In view of

(i)   the need to take immediate steps towards the creation of a rational and equitable international economic order which has as one of its primary objects the redistribution of the world's wealth in favour of the poorer countries; and

(ii)   the complexity, range and inter-related nature of the issues involved —

Commonwealth Heads of Government are urged to agree that a small group of eminent individuals selected from the Commonwealth on the basis of their personal capacities and their expert knowledge of contemporary problems of international economic development, be invited to identify measures of a practical nature which are amenable to effective implementation.

While bearing in mind the emphasis on personal expertise, the Group should, as far as possible, be assembled in a way which would enable the perspectives of the different regions of the Commonwealth and national development strategies to be brought to bear on problems.

The Expert Group should work within the framework of —

(i)   the Declarations and Programmes of Action elaborated by a number of recent international conferences and which provide a detailed picture of a just and harmonious order and in particular of the Declaration and the Programme of Action on the Establishment of a New International Economic Order adopted by the General Assembly at its Sixth Session in April, 1974; and

(ii)   the relevant principles of the Commonwealth Declaration adopted in Singapore in 1971.

Although it is not within the character of our forum to become involved in technical discussions, it is considered that we should, if this proposal is adopted, provide the Group with broad guide-lines.

The Group may, for example, be asked to pay particular attention to the ideas and proposals advanced during this debate here in Kingston as well as the following:

(i)   measures to strengthen international co-operation in the field of international trade in primary commodities within the context of the integrated commodities programme of the UNCTAD and current proposals for indexation;

(ii)   new measures which the international community can introduce for assisting developing countries with:

(a)   increasing food production;

(b)   the promotion of co-operation between themselves at sub-regional, intra-regional and inter-regional levels;

(iii)   mechanisms for increasing the flow of long-term developing funds and the transfer of real resources to developing countries, due regard being paid to the special needs of the least developed land-locked and island developing States;

(iv)   relevant concepts and mechanisms embodied in recent econo-
        mic co-operation agreements between developed and develop-
        ing countries;

(v)   institutional reform with particular reference to increasing
        the share of the developing countries in decision-making in
        the major financial institutions.

While high level expertise is mobilised in this way with the objective
of identifying immediately practical steps, it will also be necessary to
agree on some fully representative forum where the steps so identified
may be accepted for translation into action.

The Expert Group may, therefore, be asked to pay special attention
to this matter, bearing in mind the meeting of Commonwealth Finance
Ministers to be held later this year as well as the forthcoming Special
Session of the General Assembly to be devoted to Development and
International Economic Co-operation.

It would obviously be useful if the Report of the Experts or, more
realistically perhaps, a preliminary report, were to be ready by a date
that would allow Commonwealth Governments, should they so desire,
to take its contents into action when participating in the Special Session.

This proposal implies, of course, no endorsement of any set of proposals, but it provides a mechanism for examining a wide variety of
them. To such machinery, I suggest, the British Prime Minister may
well wish to refer for consideration, those proposals he has outlined to
us today.

A Commonwealth Consensus

The issues confronting mankind are planetary in their range. Whether it
be the abolition of hunger and poverty or the management of the
resources of the sea or the environment or of devising a new monetary
system, our particular moment in history is also characterized by divisions and discontinuities, by preoccupations with the interests of regions
and ethnic groupings, with the preservation of ideologies or systems.

But this Commonwealth Conference enables us to take a second look
at urgent global issues in a situation less subject to such pressures and
tugs.

The Commonwealth is already an important instrument of functional
co-operation. Here in Kingston, if we can harness realism to global objectives, we can make of the Commonwealth a consensus and bridge-building
mechanism which can significantly advance the objectives of the international society.

THE UNCTAD COMMODITIES PLAN

Introduction

The United Nations Conference on Trade and Development has long had a reputation for being more sympathetic than most UN agencies towards the aspirations of less developed countries. This is not surprising in view of the origins of UNCTAD and neither is it surprising that UNCTAD should play an important role in attempts to restructure world trade following the advent of producer power.

Following an intense period of activity on matters of raw materials supply by the United Nations Organisation during 1974, UNCTAD produced an Integrated Programme for Commodities for a meeting of the Commodities Committee of the UNCTAD Trade and Development Board in February 1975.

This plan was designed to help stabilise the markets for a large number of important commodities exported by less developed countries. It required a considerable amount of funding, estimated at about $11,000 million. The plan underwent a number of modifications within a few months of its original inception, but it is the original plan which is given here, representing as it did a plan for an advanced degree of cooperation between developed and less developed countries on the matter of resources and international development.

AN INTEGRATED PROGRAMME FOR COMMODITIES

Introduction

1. Developments in the world economy in 1974 and those now more clearly foreseen for 1975 give added force to the sense of urgency expressed by the Trade and Development Board at the first part of its fourteenth session in calling for a new approach to international commodity problems and policies (resolution 124 (XIV)). The emphasis that now needs to be placed without delay on an international approach to problems of food and raw materials is being shaped by world events. Most important, it is now evident that a serious slowdown is occurring in economic activity in the major industrial countries, accompanied by unabated inflation and radical change in the international payments situation. The transmission of adverse effects to the economies of primary producing countries must be prevented by policies that maintain over-all levels of effective demand and prices for exports. International action on commodities must be prepared to deal with this prospect. Secondly, the imbalances in commodity supply and demand since 1972, together with oil developments and the food crisis, have stimulated a fundamental rethinking of the features of an international trading system that will assure vital supplies to importers, and give adequate incentives to primary producers. This concern is heightened by the extent of the recent upward fluctuations in commodity prices, and the important initiatives taken by producing countries in these circumstances to redress the balance of bargaining power in trade through co-operative association. An integrated commodity programme must incorporate better assurances as to supplies and markets, and greater price stability at levels that are adequate for producers and equitable to consumers; it should also allow for the constructive organization of producers in order to influence the

operation of the institutional framework in international trade relations. Thirdly, present difficulties are symptomatic of long-term structural problems in developing economies brought about by concentration and over-dependence on primary production and export. An integrated programme ought therefore not to impede but should rather encourage diversification in agriculture and diversification (especially vertical) in the economy in general, based on fuller co-operation between industrial countries and the primary producing countries in their general trade policies.

2. For the first time in many years, the world is without adequate reserve stocks of essential foods and several industrial materials. The consequence may be greater instability in the world economy and in commodity trade than in the past. Yet roughly two dozen of the more important commodities or groups of commodities in world trade represent two-thirds of the economic activity for export of the developing countries (excluding oil), and over half of the external purchasing power of the national product of at least 60 countries. Unless there is radical change of approach, there is little assurance that future output trends will be able to prevent a recurrence of the present crisis, or will be adequate for the requirements of world trade or the economic development of most of the world's population.

3. On the basis of these general considerations, the following proposals are made on the key issues that would form the core of an international approach to commodity problems:

(a) Establishment of international stocks of commodities, on a scale sufficient to provide assurance of disposal of production undertaken on the basis of a realistic assessment of consumption, as well as assurance of adequate supplies at all times for importing countries, and also large enough to ensure that excessive movements in prices — either upward or downward — can be prevented by market intervention;

(b) The creation of a common fund for the financing of international stocks, on terms and conditions that would attract to the fund investment of international capital, including the support of international financial institutions, while also reflecting in its composition the responsibility of governments of trading countries for the management of international commodity policies;

(c) The building up of systems of multilateral commitments on individual commodities, whereby governments, on the basis of a multilateral appraisal of trade requirements, enter into purchase and supply commitments as a means of improving the predictability of trade requirements and encouraging

rational levels of investment of resources in commodity production. The functioning of the system, and the capacity of
governments to undertake commitments on behalf of their
export and import sectors, would be facilitated by arrangements for linking the commitments to the operation of international stocking mechanisms, and to compensatory schemes;

(d)   Improved compensatory arrangements in situations of fluctuation in commodity prices and earnings for which international
stocking or other arrangements could not secure suitable
price and production incentives;

(e)   The implementation of measures removing discrimination in
trade against processed products, encouraging the transfer of
technology and supporting a more intensive research effort,
in order to secure rapid development in the processing of raw
materials in producing countries as a basis for the expansion
and diversification of export earnings.

Chapter I

CONCEPT AND MEANING OF AN INTEGRATED PROGRAMME

4.   In his note on an over-all integrated programme for commodities
(TD/B/498), submitted to the Trade and Development Board at
the first part of its fourteenth session, the Secretary-General of
UNCTAD presented the case for a new initiative in international
commodity policy which might take the form of a multilateral
negotiation, or "package deal", for the establishment of international arrangements covering a comprehensive range of commodities of
export interest to developing countries. It was suggested that these
arrangements might be based on a common set of principles, objectives and techniques, and should be "multi-dimensional" in scope,
covering not only pricing policy but also aspects such as marketing,
diversification and access to markets.

5.   Document TD/B/498 was a response to the call made in the Programme of Action on the Establishment of a New International
Economic Order,[1] adopted by the General Assembly at its sixth
special session, for the preparation of an over-all integrated programme for a comprehensive range of commodities of export interest
to developing countries, setting out guidelines and taking into
account the current work in this field. The call for such a programme was itself the result of the current crisis of international
commodity policy, constituted by the fact that years of studies,
discussions and consultations in various forums have succeeded in

establishing international arrangements for only a few commodities, hardly any of which have proved to be effective or durable.

6. The proposal for an over-all integrated programme for commodities endeavours to launch international commodity policy onto a new course which, it is hoped, may have a greater chance of success than the approaches hitherto adopted. The proposed new approach is an attempt to move urgently from the field of consultation to the field of negotiation. To facilitate this shift, it is proposed that arrangements for a comprehensive range of commodities should be negotiated in the form of a package, so that the special interest of countries in some commodities could be an incentive to them to reach agreement on others. Although this implies a departure from the traditional piecemeal, commodity-by-commodity, approach to negotiations, it does not alter the fact that specific arrangements would have to be devised for individual commodities. It does mean, however, that the drawing up of arrangements for a substantial number of commodities would have to be agreed upon at the same time and undertaken simultaneously, or as simultaneously, as possible, this being an important dynamic feature of the new approach.

7. Fundamental to the proposed new approach is the setting of wider objectives for international commodity arrangements, including the improvement of marketing systems, diversification (horizontal and vertical), expanded access to markets, and measures to counter inflation, in addition to the traditional objectives of stable and remunerative prices. Acceptance of these additional objectives is essential if more viable and more durable commodity arrangements are to be established. Without some kind of provision for relating prices of exports to prices of imports[2], for example, commodity arrangements operating in conditions of rapid international inflation would tend to break down. In the case of some commodities, provision for improved access to markets would also be essential, since the expansion of world supply in line with demand cannot be assured if efficient producing countries fear that any attempt to expand their exports will be frustrated by import restrictions, as has been notably the case with sugar and livestock products.

8. The proposed new approach recognizes the principle, as reflected in the Havana Charter, of co-operation between exporters and importers, or producers and consumers. But it also recognizes that co-operative action by producers can be — and in given instances has been — an aid to negotiations with consumers. Indeed, in some cases, such co-operation might be indispensable to negotiations and generally it would be an important means of expediting

and facilitating progress. It should therefore be accepted and encouraged.

9. One final point needs to be made regarding the relationship between the proposed integrated programme and the work of existing international commodity councils and consultative groups. In document TD/B/498, it is stressed that these bodies should be consulted, and that their work should be taken fully into account, in the preparation of the integrated programme. Some of these bodies, in fact, already plan to draw up and negotiate economic arrangements for the commodities with which they are concerned. Such work could be perfectly compatible with the idea of the integrated programme, and indeed could form part of it, provided the governments concerned carried out the work in the manner and in the spirit of the programme.

10. Briefly stated, the principles and objectives on which it is proposed the programme should be based are the following:

(a) There is a need to seek solutions simultaneously and urgently to the problems of a number of commodities of major interest to developing countries, both as exporters and importers, in view of the considerable threat to the interest of these countries posed by prospective developments in the world economic situation in both the short term and the longer term (see paragraphs 1—2 above):

(b) International action on commodities should take due account of the interests of both exporting and importing countries;

(c) Co-operative action by producing countries has a legitimate and important role to play in solving the problems of individual commodities;

(d) Arrangements for the stabilization of prices, in the sense of the smoothing out of irregular or cyclical fluctuations, are required for many commodities in order to allow correct responses to price incentives in production, to help stabilize export incomes and import bills, and to improve the competitive position of natural raw materials facing competition from synthetics;

(e) Commodity prices should be at levels which provide incentives for the maintenance of adequate levels of production, which are just and remunerative, which take due account of world inflation, and which are consistent with developmental objectives;[3]

(f) Diversification and the expansion of processing of commodities in developing countries should be encouraged;

(g) The improvement of marketing and distribution systems, more advanced technology, and research should be actively promoted in the interest of both importers and exporters;[4]

(h) International commodity arrangements should seek to ensure liberal access to protected markets for exporting countries and security of supplies for importing countries.

Provisions for attaining additional agreed objectives could be incorporated in arrangements for individual commodities or groups of commodities in the course of detailed negotiations on the programme.

Chapter II

ELEMENTS OF THE PROGRAMME PROPOSED

A. International stocking policies

11. The most urgent need is for action on commodity stocks, and it is suggested that this be given priority consideration. The urgency is dictated by the recent and possible future developments in the world economy. Prices of a wide range of primary products have been falling precipitously in recent months, despite continuing international inflation. While declines from the 1973–1974 peaks were to be expected, their extent and speed have been ominous. By November 1974, prices of copper, rubber, zinc and wool had fallen by more than 50 per cent, cotton and tropical vegetable oils by 30–50 per cent, and iron ore, lead, abaca and tin by 20–30 per cent. A continuing shortage of some agricultural products coupled with high prices, notably for sugar, in the midst of the present recession, can be attributed in large part to past under-investment associated with long periods of low prices. To wait until the downturn becomes more general may be leaving matters too late to negotiate on corrective measures of international scope. It is necessary to agree on plans now to prevent over-reaction and slump.

12. It is proposed that international stocks should be established for a wide range of commodities by purchasing them when their prices are at an agreed floor level. Table 1 lists 18 major commodities which appear suitable for international stocking. The list is provisional. Some of these commodities, for example, wheat and wool, are exported predominantly by the developed countries, but are important to developing countries as both exporters and importers. Some others, for example, tropical beverages and natural rubber, are produced exclusively in the developing countries. The 18 commodities listed account for 55–60 per cent of the total primary

product exports of developing countries other than petroleum, and their stabilization at an adequate level would have far-reaching effects on these countries.

13. The accumulated stocks would serve as an international reserve of foodstuffs and industrial raw materials which would help to assure an uninterrupted flow of world consumption and world industrial production. They would be released to the market or to the participating countries when prices moved above an agreed ceiling. Such international reserves should be created, since the national stocks of some key products, very large until several years ago, have now been depleted. By the end of 1973, the aggregate of existing stocks of cereals and most non-ferrous metals had fallen to under 10 per cent of world annual consumption, and of several other major foodstuffs and agricultural raw materials to under 25 per cent. Most of these stocks do not leave any reserve margin above working ("pipeline") stock requirements. Stocks will have to be rebuilt, although not necessarily to earlier levels. Since it is unlikely that any one single country will attempt in the future to hold stocks for the world economy, the question is whether there will be attempts by a number of countries to carry stocks individually, or whether there will be an international system of stock accumulation, holding, and disposal.

14. The present proposal is not limited to meeting the current emergency situation. The machinery of international stocks, once created, should remain in existence and would then be able to exercise a continuing stabilizing effect on world commodity markets, in the interest of both the exporters and the importers. In the absence of price support by an international stock, the exporting countries, particularly low-income ones, are frequently compelled, in periods of excess supply or weak demand marked by falling prices, to sell on a declining market, thus depressing prices and earnings even further, because they do not have enough financial resources to hold back supplies. While the support provided by an international stock may not be sufficient for them to achieve an adequate price level and other measures may be needed, such support is likely to be a necessary condition in most cases. For the importing countries, international stocks would bring security of supplies and reasonably stable prices; also, by providing producers with certainty concerning prices and markets, such stocks would help assure adequacy of supplies for the importing countries over the long run, especially if supported by investment on the part of international development agencies designed to expand output of scarce materials. A wide geographic distribution of physical location of stocks would be an additional guarantee of equal access to primary products.

Table 1. Major stockable commodities: Trade values, 1972[a]

(Millions of United States dollars)

	Exports f.o.b.				Imports c.i.f.			
	World	Developed market-economy countries	Socialist countries	Developing countries	World	Developed market-economy countries	Socialist countries	Developing countries
Wheat[b]	4 366	3 818	388	160	4 609	1 540	1 291	1 778
Maize	2 298	1 914	53	331	2 444	1 905	324	215
Rice	1 120	537	143	440	1 232	175	82	974
Sugar	3 334	921	178	2 235	3 379	2 304	460	614
Coffee (raw)	3 049	—	—	3 049	3 368	3 101	126	141
Cocoa beans	723	—	—	723	729	572	131	26
Tea	745	79	57	609	784	470	72	242
Cotton	2 828	587	484	1 757	3 055	1 714	792	519
Jute and manu- factures	762	71	21	670	840	520	120	200
Wool	1 346	1 143	42	161	1 722	1 361	257	105
Hard fibres	87	3	—	84[c]	106	92	7	7
Rubber	904	—	—	904	1 095	689	305	101
Copper	4 113	1 364	354	2 395	4 226	3 635	377	214
Lead	418	257	45	116	470	379	60	31
Zinc	862	558	110	194	938	736	77	125
Tin	730	70	28	632	758	613	53	92
Bauxite	305	82	5	218	363	325	36	2
Alumina	609	265	46	298	685	532	91	62
Iron ore	2 608	1 213	403	992	3 484	3 039	425	21
Total	31 207	12 882	2 357	15 968	34 287	23 702	5 086	5 499

Source: FAO, Trade Yearbook 1972, and national statistics.

[a] The figures are preliminary. In sugar, cocoa and copper, import values appear understated in relation to exports. In metals and ores, EEC intra-trade is excluded.

[b] Including flour.

[c] In addition, $49 million exports of hard fibres manufactures.

Notes: Oilseeds and vegetable oils are under consideration. Special investigation is needed of a suitable stabilizing mechanism for oilseeds and vegetable oils produced in the tropical areas, which are interchangeable with other oils and fats in a varying number of uses

15. While the primary objectives of the proposed international stocking
     system are price stabilization at an adequate level, and assurance of
     supply and outlets, it should also aim at a financial profit on its
     operations as a whole. In this way it would be able to discharge its
     economic functions in a sustained manner. The buying and selling
     prices should be subject to re-examination at regular intervals, ini-
     tially perhaps once a year, in the light of experience, particularly
     with respect to the level of purchases and sales by the stock. If in
     serious surplus situations it became necessary for the producing
     countries to introduce temporary mandatory restrictions on exports,
     it is suggested that such restrictions should be made proportional
     to output at the time of introduction, thus avoiding the need for
     extended quota negotiations in advance as well as the danger of
     freezing the geographic pattern of production. Since the lasting
     solution to persistent cases of over-production is structural readjust-
     ment, it is proposed that the incentives provided by the operations
     of an international stock should be used to stimulate resource
     shifts and resource mobilization policies designed to accelerate
     diversification of production and exports.

16. The cost of acquisition of the necessary volume of the 18 commo-
     dities listed in table 1 has been provisionally estimated at $US 10.7
     billion, assuming the commodities were bought at average prices
     prevailing in the five-year period 1970–1974. Of this amount,
     $4.7 billion is accounted for by grains (wheat, rice and coarse
     grains). The next largest amounts are for sugar, coffee and copper,
     aggregating $3.2 billion. If the commodities were bought at average
     prices of the three-year period 1972–1974, the aggregate cost
     would be one-fourth higher, and at 1970–1972 prices probably
     one-fourth lower. Applying prices prior to 1970 would not be use-
     ful for these illustrative estimates, in view of the distorting effect
     on all nominal values of international inflation. The estimate for
     grains assumes a reserve stock based on the FAO analysis of world
     needs prepared for the World Food Conference; international grains
     stocking requirements would be lower if they were designed pri-
     marily to assure supplies for the low-income countries (mostly in
     South Asia and Africa) and for other developing food-importing
     countries (i.e. some petroleum producers). Conversely, the esti-
     mated stocking requirements for some industrial raw materials
     may well be on the low side.

17. The aggregate figure of $10.7 billion represents conceptually a
     commitment rather than a disbursement figure, for two reasons.
     First, the estimates for each commodity represent the maximum
     which may be accumulated; in practice, the accumulation will vary
     between zero and the maximum. Secondly, if a multi-commodity
     stock or a common financing fund is in operation, purchases of

some commodities will be offset by sales of others in particular periods. The extent of the offset will be determined by the amplitude of the international business cycle: the stronger the cyclical movements, the more likely that most commodities will move together and there will be little offset. During the post-war period, some commodity prices moved in opposite directions to others in most years; but if the future is likely to show more cyclical instability than the last two decades, the offsetting action will be weaker. The policy implication of these factors is that access to resources should be as large as possible, although the actual use of funds in particular periods may be below gross commitment needs. Further work is needed on the selection of commodities, size of stocks, price ranges, and likely length of the stocking periods, in order to improve the estimates given above.

18.   International stocking mechanisms exist at present for only two of the 18 commodities, tin and cocoa. The operating experience and financial results of the tin stock, covering the last 20 years, have been favourable. The mechanism of the cocoa stock has just been established. The limited use of stocking arrangements in international commodity policy is inconsistent with the present and prospective needs of the world economy.

19.   The functions of an international stock can be performed by national stocks in the producing and consuming countries, provided the producers have sufficient finance, and all national stocking policies are internationally agreed. The first condition is met by some producing countries but not the majority. The second condition has not been realized in practice, although it is feasible in principle; it would require institutional arrangements and a set of agreed rules concerning the range within which the participating governments would try to keep the world market price, and the stock-releasing and stock-accumulating obligations of each government. Even if these conditions were met, the aggregate of national stocks, each based on individually perceived country needs, could be expected to be larger than an internationally managed stock which could be deployed more efficiently to achieve the same objectives: the latter would have to be of a size to cover the net deficit (or to absorb the net surplus) of world production in relation to consumption, while the aggregate of national stocks would tend to be of a size to cover the gross deficit, i.e. the sum of deficits of individual countries. In addition, there would be an element of uncertainty as to access to supplies since the stocks would be nationally owned and operated. Some of these problems could be partly resolved by the simultaneous existence of an internationally managed stock and national stocks, all operating in a co-ordinated manner. Such co-ordination would require a sustained effort in

international co-operation. Nevertheless, if the traditional policies of the leading importing countries towards international stocks for commodity stabilization are maintained, it will be in the interest of producers, faced with stagnation or worse in their economies, to adopt stocking arrangements among themselves.

20. Early action on international commodity stocks is seen as the cornerstone of the integrated programme, and the question of organizational arrangements will be a first consideration. The proposed programme embraces the establishment of international stocking arrangements for most commodities of significance to developing countries in their capacity as substantial importers or exporters. In some cases, such arrangements might be organized as a series of individual commodity schemes, in which the stocking operations could be one among several types of measure in a multi-dimensional approach to the short-term and long-term adjustments required. The common fund for the financing of stocking schemes, also proposed in the programme, would help to ensure consistency in the objectives and conduct of individual commodity schemes.

21. On the other hand, it will be of little real value to engage in arrangements for financing international stocks if slow progress is made with the arrangements for the stocking operations themselves. It may be necessary to organize generalized arrangements for stocking those commodities on which progress could not be made in individual commodity schemes despite a consensus on the problems. Moreover, the organization of stocking arrangements would have to be viewed in a different perspective if the onset of a serious general recession led to the prospect of a major decline in the volume of trade and prices for export commodities of cardinal importance for developing countries, while prices of basic foods and general inflation rates remained high. While early action on action on stocking arrangements is in any case required, such conditions would make it all the more imperative to set up a rapidly organized scheme to acquire stocks of a number of commodities. Such a scheme might have to be organized on multi-commodity lines, under a single agency, and in large part to provide an outlet for production already initiated. It might have to be set up as a transitional measure, allowing time for the establishment of more comprehensive measures under an integrated programme.

22. There are other reasons for a multi-commodity approach, which would either co-ordinate stocking for a group of related commodities, or would more ambitiously seek to manage stocks of a comprehensive range of commodities. The functions of a central agency could have the advantage of incorporating the purposes of a common financing fund, as well as those of stock management,

though it would be necessary to consider the extent of its involvement in other types of commodity adjustment policies, such as supply management. The functions of a multi-commodity organization set up with the wider objective of co-ordinating the activities of a number of commodity agencies could include longer-term objectives. It could be responsible for, and be capable of, taking into account the effects of policies for one commodity on others. It could give guidance to individual commodity agencies on diversification policies, and it could set a better perspective for long-term planning on crops with long gestation periods than could individual agencies of medium-term duration. This type of operation could also, as noted above, make a significant contribution to the control of the business cycle, by co-ordinating the purchases of commodities in the downswing to support effective demand and protect employment, and by releases of commodities in the periods of upswing and inflationary pressures.

23. Commodity stock operation could also be integrally linked with multilateral commitments on trade, as advocated in paragraph 38 below. The support of stocking arrangements would be a strong inducement for governments to enter into supply and purchase commitments, while at the same time such commitments should make for greater smoothness of operations by stocking agencies, by helping to prevent the need for abrupt adjustment in prices or production to divergent trends in supply or demand. In brief, governments that entered commitments would be entitled to the use of stocking facilities, as exporters or importers, with respect to non-fulfilment of the commitments. Policies would need to be elaborated with regard to the role of stocks in this manner, and these policies, combined with those on multilateral commitments, could be mutually reinforcing in encouraging trends towards global equilibrium in national supply and demand.

B.   The financing of stocks

24. The integrated programme will need to adopt a broad solution to the financing of stocks as a key element in the programme. The illustrative example given in the study on the role of international commodity stocks (TD/B/C.1/166/Supp.1) suggests that the comprehensive programme envisaged might involve capital resources of the order of $11 billion (nearly half of which might be for grains alone), though the amount in use at any time might not approach this size, and would in part represent re-allocation of public expenditures already committed to stocking (see para. 17 above). International co-operation in investment of this magnitude cannot there-

the approximate amounts of a commodity that each government expected the economy to supply or demand, too or from all participating countries, based on forecasts of trade, including state trade, private and official bilateral contracts and open market trading. Though the agreement would carry an obligation, as described below, it would not have the full character of a contract obliging governments as such to acquire or supply amounts that were not purchased or sold under the ordinary trading system of the country during the period of the commitment.[5]

33. Assurance of supply is the second outstanding area of concern emerging from the chaotic commodity developments of the 1970s. Set alongside the chronic concern of primary producing countries with sustained market capacity, there is a unique opportunity for a marriage of interests in realizing more predictable and more stable movement of commodities in international trade. Governments now seem more ready to recognize that reciprocal trade volume commitments could facilitate forward planning of resource use in their domestic economic policies and in solutions for balance-of-payments difficulties. These commitments could best be achieved multilaterally, and with as wide a coverage of trade flows for particular commodities as possible.

34. Trade arrangements that could offer some guidelines for national trade operations and for production planning would be valuable on several grounds. Most generally, national authorities dealing with foreign exchange budgets, import policy and the expansion of exports have to take decisions based on appraisal of market developments and the probability of national actions affecting their trade. The market price mechanism and such national actions, as short- and medium-term indicators, cause serious difficulties for such planning.

35. An additional advantage of governmental purchase or supply commitments organized multilaterally, is that they might improve terms and conditions in the extensive proportion of commodity trade that is already transacted through forward contracts. Conclusion of the commitments multilaterally would help to bring more complete information and competition to the conclusion of bilateral private or governmental contracts, especially for commodities for which open market prices are not available or representative. One of the purposes of producers' associations is to improve their bargaining position in this respect. The commitment procedure, involving as it would the exchange of information with other producers, could be of considerable value to exporting countries as regards their sales policy and the regulation of foreign companies operating in their territories.

36. The difficulty in making commitments on trade is that the influence that governments can exercise on the performance of producing and trading sectors varies a great deal. The nature of agriculture does not ensure export availability, and most importing governments cannot enforce purchases by commerce and industry. Moreover, even if such commitments were undertaken for a period of as short as one year, the ability to project requirements and supplies differs between commodities. Nevertheless, it should be possible to proceed on the basis of the best possible short-term forecasts of import demand and export availabilities, undertaken by national authorities in consultation with the private interests concerned.

37. To be effective, a system of multilateral commitments would need to cover a large proportion of trade in any given commodity. Otherwise, any significant imbalance in import demand and export supply might not be discernible. Producers' associations could find an important place in the system, by determining the export potential and price conditions acceptable to their membership as a basis for a constructive relationship with consuming countries. Bilateral trade arrangements and state trading practices would also fit well into the procedures of multilateral commitment, since such transactions assist governments in deciding on their over-all requirements and supply.

38. The likelihood of governments being able to contemplate commitments would be enhanced in these circumstances if a supporting role were to be assumed by international stocks, whereby the obligations of governments to fulfil commitments that had been realistically appraised but impeded by circumstance would be mainly taken over by the management of stocks. The use of stocking arrangements in this way would need to be symmetrical, assisting both countries whose commercial performance left a deficit in import commitments as well as countries whose export availabilities failed to reach the amount committed. One method of operation could be to allocate shares in an international stock to participating countries that could be taken up against non-fulfilment of commitments. Such stocking rights could be transferable. While the financing of stocks might be independently provided through a common fund, the use of stocking rights in this manner, by either exporters or importers, could entail certain charges (possibly determined in relation to carrying costs) as a disincentive to deliberate evasion of purchase or supply commitments. Such charges would provide a useful source of operating revenue for stocking agencies, but would not place a large financial penalty on commitment deficits. As noted in para. 23 above, pricing policy would need to be

harmonized under stocking arrangements and in the determination of multilateral commitments.

39. Multilateral commitments could also be concluded on commodities for which stocking arrangements were not feasible, especially if compensatory schemes were also available to countries with a high dependence on non-storable export products. There may also be scope for arrangements that balance commitments on one commodity against commitments on another.

40. The system could thus be envisaged as a three-stage process applying to trade in one or several commodities as follows:

 (i) Projection of the global potential for trade between exporters and importers over a specified period, at least annually and preferably over the medium term, as a forward planning exercise;

 (ii) Consultations between producers (or consumers) as a means of resolving significant coverages in annual export availability or in import demand, the range of prices applicable, or other terms of arrangements;

 (iii) Agreement in the form of purchase and supply commitments concluded multilaterally, but without specification as to the direction of trade.

41. The question arises whether stocking arrangements will by themselves perform many of the functions claimed for multilateral commitments. A system of international stocks alone can realize important objectives, including the reduction of price fluctuations and the effecting of short-term adjustment for imbalances in production or consumption in a manner less abrupt than results from the operation of market forces and prices alone. In this respect, stocking arrangements will help to reduce the uncertainties in planning trade and production policies. However, as stressed in section A above on international stocking policies, when they are not open-ended they will usually need to be associated with provisions for quantitative controls and alteration of intervention prices if and when stock operations are not achieving longer term equilibrium in supply and demand. Such measures effect adjustment when the problem has developed and place the burden on export producers; forward commitments could encourage a less painful or less wasteful adjustment process. While, therefore, stocking arrangements would not need to be associated with multilateral commitments in order to perform a central role in an integrated programme, their value might be more effectively realized in dealing with temporary and irregular market disturbance, particularly from the demand side, if price movements and quantitative restrictions due to struc-

tural imbalance could be averted through forward trade commitments. On the other hand, a multilateral commitment system would most probably stand a greater chance of acceptance and would be greatly strengthened in practice if supported by international stocking arrangements.

D.   Compensatory financing of commodity trade

42.   The foregoing measures would still leave certain countries vulnerable to export instability and to depressed trends that reflected a weak bargaining position for key exports. These would probably be countries with a significant dependence on commodities for which stocking arrangements or multilateral commitments are not feasible, or where participation in a multilateral commitment still left serious fluctuation in their prices or returns. Such commodities are less likely to have a significant influence on the trade or payments situation of importing countries, so that arrangements for compensation could be more suitably directed to the assistance of exporting countries.

43.   The justification for such assistance is already concretely acknowledged in the international community through the IMF facility for the compensatory financing of export fluctuations, and by a major group of importing countries in the EEC proposals for a commodity compensation scheme available to associated and associable countries. Medium-term loans by the IMF are intended to smooth out fluctuations in the total export trade of primary producing members and thus compensate for fluctuations in specific commodity exports when these are reflected in downturns in overall export receipts. The IMF scheme is self-financing, with full repayment of loans. It does not lay down conditions with regard to the domestic arrangements the financial authorities of borrowers make on producer prices or incomes. About SDR 1 billion of assistance was provided to 32 countries in 1963–1973. The EEC scheme would be applicable to a selected group of commodities, and while based on returns to those commodity sectors individually, is intended to have an income-stabilizing influence for the individual producer in recipient countries. It has provisions that would cause non-repayable expenditures, though partly self-financing.

44.   For those countries in the position described in paragraph 42 above, the expansion and liberalization of IMF assistance would probably be of significant help. The introduction of the EEC scheme would also be helpful in finding solutions to this kind of problem. There would probably continue to be problems in some special areas of commodity trade for particular countries that it might not be pos-

sible to encompass in the IMF approach at the level of total export earnings, but the extent of these problems would probably be greatly reduced if appropriate changes could be made in the IMF facility to enlarge the amounts transacted and the number of countries using the facility. Within an integrated approach to commodity problems, it would seem best to give priority attention with regard to compensation aspects to the possibility of building on the present IMF facility, with attention being given at a later stage to any additional compensatory measures that might be required in consideration of the scope of the expanded facility.

45. The aspects of the facility in mind in this respect are (i) the need for more flexible conditions as regards the balance-of-payments criterion for assistance; (ii) relaxation of the limits on the amounts available as determined by IMF quotas to take account of the size of shortfalls; (iii) easier requirements on the completion of detailed export statistics within a relatively short period of the shortfall in exports; (iv) extension of the repayment period beyond the present obligation to make complete repayment within five years, including a closer link with the recovery of exports; and (v) account to be taken of changes in the import purchasing power of a country's exports.

46. If, however, action on these lines did not appear feasible, the problem for the exporters of the perishable commodities for which demand and prices are highly unstable would remain large enough to warrant consideration of commodity compensation schemes. The most critical point of policy in a commodity compensation scheme is that the scheme should be considered as a residual measure, to be applied when other more direct approaches are inappropriate or inadequate to meet the ultimate objective of stabilizing and maintaining the real export income of exporting developing countries. Consequently, the scheme should be designed to provide automatic compensation payments in the form of loans to developing countries experiencing shortfalls in their export income from the commodities considered (i.e. those not covered by other arrangements). Such loans would be repaid out of part of any excess of exports over the agreed "normal" levels (from which the shortfalls were calculated). Repayment procedures might also include provision for conversion of unpaid balances into grants.

E.   Expansion of processing and diversification

47. The contribution of commodity production and trade to the economic development of the developing world will only be realized rapidly and efficiently if shifts in resources within the primary

sectors of the developing countries and within their economies in general can occur. The measures above would create more favourable conditions for appropriate diversification and for freeing resources in a more broadly based economic structure. But in addition, separate and more constructive attention will be required in the international community to develop means of expanding the processing of primary products, removing trade discrimination in this respect, and to encourage the transfer of technology and research with this objective.

48. For the generality of primary commodities exported by developing countries, there is need for greater diversification into the more manufactured forms of the basic products. Expansion of trade in semi-processed and processed products is inhibited by various factors, among which are tariff and non-tariff barriers in developed countries. This situation needs to be improved through extension of GSP coverage to more products of primary origin, the removal of ceilings and quotas under the GSP, and the removal or relaxation of other non-tariff barriers. In addition, increased attention needs to be paid to the problems of the development of processing in the economies of developing countries, including such aspects as the transfer of technology and research. Among the actions that developing countries can take, will be important to provide export incentives.

Chapter III

APPLICATION IN PRACTICE OF THE VARIOUS PRINCIPLES, OBJECTIVES AND TECHNIQUES

49. In considering how the various principles, objectives and techniques dealt with in the preceding chapters might be applied in practice, four separate groups of commodities which face distinctive basic problems and for which distinctive policy objectives need to be set can be taken as a basis for negotiation. This illustrative grouping is not meant to represent any exhaustive list of commodities, nor to be exclusive regarding the techniques indicated. Further, it is without prejudice to the possibility of adopting a multi-commodity system and arrangement of the type referred to in paragraphs 21 and 22 above and in the supporting document on international stocks[6] or the proposal contained in paragraph 51 below.

50. The first group that can be distinguished consists of *essential foodstuffs,* in respect of which the prevention of excessive fluctuations

in world prices in both the short and long term, the provision of adequate incentives to production, and the building up of large internationally-held reserves, is in the common and vital interest of virtually all countries of the world, exporting and importing, developed and developing. Because production of these commodities is widespread throughout the world and subject to unpredictable fluctuations, supplies and prices at the world level can be effectively administered only by means of stocking arrangements. For the reasons described in the supporting paper on international stocks[7], the most economic and effective way to achieve the required degree of supply and price stabilization, and the necessary margin of supply security, would be to establish internationally owned and internationally managed buffer stocks, with financial support from the proposed common fund. The current world shortages of cereals and sugar have underlined the fact that national action, far less private commercial action, cannot be relied upon to ensure adequate prices and security of supply for basic food commodities, shortages of which can cause hardship, suffering or death to millions of human beings. Indeed, international stocks should be established as a matter of urgency for *wheat, coarse grains, rice, sugar* and *selected vegetable oilseeds and oils*[8]. The international co-ordination of national stocks, as proposed by the World Food Conference, would be a major improvement in the organization of world trade in food, though it should be regarded as an important first step towards the establishment of international stocks.

51. In view of the inter-relationships between markets for different cereals, and those between the markets for different vegetable oilseeds and oils, it would be necessary for a single, multi-commodity, stock to be established for each of these two groups of commodities[9]. In order to provide further safeguards for importing and exporting countries, stock arrangements could be reinforced by commitments by exporting countries regarding the flow and allocation of supplies in cases of unexpectedly severe shortages, and by commitments by importing countries regarding *minimum degrees of access to protected markets*, along the lines suggested in the supporting paper on the role of multilateral commitments in commodity trade (TD/B/C.1/166/Supp.3).

52. The second group of commodities consists of *essential industrial minerals subject to substantial price instability and/or unsatisfactory price trends*. These commodities include bauxite, iron ore, copper, lead, zinc, manganese ore, tungsten and phosphates (in addition to tin, for which an international buffer stock already exists). In view of their importance in the export structure of many developing countries (and some developed primary export-

ing countries), and in the import bills of many developed and developing countries (because of the inelastic nature of the demand for them), there is a large common interest for both exporting and importing countries in the prevention of excessive fluctuations in their prices. In addition, importing countries have a strong interest in the longer term security of supplies of these commodities at "reasonable" prices while exporting countries are also extremely concerned to maintain their prices at adequate levels, taking into account world inflation, the conditions of demand for the particular commodity, and the rate of depletion of reserves.

53. One way in which these objectives might be attained simultaneously would be through international stock operations of the kind in effect for tin. The price range to be defended by such a stock could be set by agreement at a level which would reconcile the price ambitions of exporters and importers, while the stock itself, provided it had substantial resources, could provide security of sales outlets for exporters and security of supplies for importers. This would represent a solution substantially the same as that proposed for basic foodstuffs.

54. However, the level of production of mineral commodities can be planned much more easily than that of foodstuffs. Hence effective management of world supplies and prices of mineral commodities can, in principle, be achieved by producing countries without the help of international stocking mechanisms, provided their production policies can be concerted sufficiently closely.

55. In taking action towards that end, producing countries would be acting in accordance with the principles of the sovereignty of all countries over the exploitation and use of their own natural resources. Nevertheless, they would need also to respect the principle that, in acting jointly, producers should take due account of the interests of consumers. Alternatively, this objective might be met by negotiation with consumers, where necessary, to provide them with a substantial degree of security of supplies and stability of prices over the short to medium term[10].

56. The third group of commodities facing distinctive problems is *agricultural raw materials*. These include cotton, natural rubber, jute, hard fibres such as sisal, wool, etc. Virtually all of these materials face strong competition from synthetic substitutes and, as a result, the terms of trade of many of them, in other words the ratio of their prices in the world market to the prices of manufactured goods in world trade, have shown a long-run deterioration. The increase in the price of oil, as a raw material

for synthetic production, but also as an input in the production of natural materials, has altered this relationship, possibly in favour of natural products. It would however, not be possible, even through concerted international supply management, to maintain the price of such products at levels which were out of line with their value in demand in relation to the course of prices for synthetic materials. The chief way in which producers of these commodities can exert a sustainable upward influence on their prices is by increasing consumer and manufacturer preferences for them through improvement of their quality, attractiveness and technical characteristics and through market promotion.

57.   Another important way in which demand for these natural commodities could be expanded is by assisting manufacturers in their planning and operations by providing assurances with regard to the availability of raw material supplies and their cost. There is evidence that the considerable instability in prices and to some extent in supply influences the choices made by manufacturers in favour of synthetic materials that are supplied at fixed prices which change relatively infrequently[11]. The basic objective which should be pursued for these commodities in the context of an over-all integrated programme, therefore, is assurance of delivery at predictable and stable prices. Since a fairly close regulation of market prices is required, stabilization should be sought by means of buffer-stock operations, and, to the extent possible, by offering natural raw materials on similar delivery and price terms as can be obtained from the manufacturer of synthetic materials. Producing countries have also expressed a strong interest in alternatives to the day-to-day open market system of trading in these commodities.

58.   The collaboration of importing countries in such operations would help greatly to ensure their success and it is to be hoped, therefore, that they would participate fully in the financing and management of stocking arrangements. However, in case importing countries show insufficient interest, exporting countries should be assured of adequate international financial support to enable them to establish effective stocks on their own.

59.   As part of an over-all integrated programme for commodities, therefore, international stocks should or could be established for *cotton, natural rubber* and *wool.* In addition, stocks should be established for *jute* and *hard fibres,* including the simply processed forms of jute (yarn, bags and cloth) and of hard fibres (cordage), since it is in these forms that the bulk of world exports of the former fibre, and a large and increasing proportion of the latter, now enter international trade. As the products concerned are of a

fairly standard and homogeneous nature, stocking operations for them should not present any undue technical difficulty.

60. The fourth and final group of commodities whose problems need to be separately distinguished is a group of *tropical beverages and fruits which have shown a tendency to over-production or which are subject to cycles of over-production and shortage*. As a result of these tendencies world prices of these products have either been depressed over a considerable period (as in the case of tea, bananas, oranges and tangerines), or subject to wide fluctuations (as in the case of cocoa and coffee). Even the recent exceptional commodity price "boom", it may be noted, failed to lift (or maintain) the prices of these commodities (except perhaps cocoa) to satisfactory levels, especially if the recent acceleration of world inflation is taken into account. It is probable that in the production of most of these commodities real wages have been falling.

61. A common feature of the situation of all these commodities is that solutions to the problems facing them depend on co-operative remedial action by producers, as well as the assistance of the international community. Action by producers is needed to manager supply through restraint on investment. The action of the international community is needed to assist in financing stocks for those products that are storable and to conclude trading commitments on those which cannot be economically stored. Special arrangements may be useful for the disposal of temporary or marginal surpluses, as in the stocking revisions of the International Cocoa Agreement.

62. For cocoa and coffee, which are marked by both long cycles and sharp short-term variations in production and prices, a special blend of stabilization policies may be needed. Short-term fluctuations can be offset by international stocking operations combined with export quotas as needed; the solution to the long cycle has to be sought in policies stabilizing the rate of investment and promoting shifts to other activities. A progressive export tax applied when prices are high would dampen the cycle as a whole, because it would curtail investments in surplus capacity and therefore avoid prolonged periods of depressed prices and earnings; in addition, if tax proceeds are used for financing diversification, there would also be income generated by the released factors of production. For tea, an agreement may be facilitated by international assistance for stocking and diversification, thus helping to remove obstacles to agreement amongst the producers concerning future output and investment. More generally, operation of an international stock could be used to stimulate desirable resource shifts and resource mobilization policies in the producing countries. In bananas, co-operation

among the producing countries in deciding on an agreed selling policy is a pre-condition for an improvement in prices.

63. The co-operation of importing countries in "policing" export quota arrangements, or in concluding multilateral commitments with exporting countries, might be of valuable assistance to the latter in their efforts to improve the stability or the level of the price of their commodity. On the other hand, exporting countries may wish to rely more on their own cohesion and self-discipline by attempting to operate such arrangements as minimum price agreements, central selling systems, uniform export taxes or co-ordinated stocking systems, according to the characteristics of the different commodities and their markets.

64. If effective arrangements for the commodities in the four groups described above could be established on the lines indicated, a major step forward would have been taken in dealing with the "commodity problem" at the world level. Inevitably, however, the extent to which the commodity problems of individual developing countries would be remedied would vary greatly, depending on the composition of the exports of the different countries and the degree of success of the various arrangements as established. For most developing countries there would be a residual problem of some degree or other, reflected in the extent to which the trend of each country's commodity export earnings, in terms of its import purchasing power, was satisfactory or otherwise. In the light of this, an *extension and improvement of the existing compensatory financing facility of the IMF* would be of value for the purpose of dealing with these residual commodity problems of individual developing countries, on the lines described in the supporting paper on compensatory financing arrangements[12]. One way of dealing more directly with the problems of individual commodity sectors would be to offer the option of obtaining assistance in relation to shortfalls in their *commodity* export earnings, rather than in relation to shortfalls in their total export earnings, if the amount of assistance obtainable under the former option would be greater than under the latter.

65. Finally, whatever the particular objectives pursued and the particular techniques adopted for individual commodities or groups of commodities, and whether or not a multi-commodity system is followed or a multi-commodity agency is instituted, the integrated approach (even in the illustrations given above) should be as comprehensive or as multi-dimensional as possible. That is, it must attempt, at least in the long run, to encompass the totality of the commodity problem from production to consumption. Above all, it must be dynamic in concept so that it allows for efficient allocation of resources, and for the structural

transformation of the economies of developing countries, especially through the processing of raw materials and foodstuffs in developing countries and the diversification of their exports.

Chapter IV

PROPOSALS FOR FURTHER ACTION

66. Paragraph 8 of resolution 124 (XIV) of the Trade and Development Board requests the Committee on Commodities to give priority consideration to the matters contained in the part of the resolution dealing with an over-all integrated programme for commodities, and to make recommendations, including a timetable of work, for appropriate action by the Board at its sixth special session.

67. In response to this request the Committee will wish to bear in mind that the integrated programme outlined in the preceding chapters proposes action on:

(a) International stocking arrangements for various commodities to be brought into operation to counter rapid deterioration of prices or downturns in demand and, in some instances, to restore dangerously low levels of world stocks. Eighteen of the principal commodities (or commodity groups) in world trade, to which these measures could be applied, have been provisionally identified;

(b) Financial support for all stocking operations through a common fund, based on contributions shared by importing and exporting countries, assisted by the international financial institutions, and also open to international investment from other sources;

(c) Multilateral purchase and supply commitments by governments to give assurance of supply and outlets on at least the key commodities in trade for which such assurance is important. These measures should as far as possible be linked to international stocking arrangements. Supply commitments are also required when independent measures are taken by exporting countries;

(d) Improved compensatory arrangements, primarily through extension of the IMF compensatory financing facility;

(e) Expansion of trade in processed products through extension of the coverage of the GSP, the removal of non-tariff barriers and the provision of export incentives.

68. While the Secretary-General will proceed with further work in the light of the comments and decisions or recommendations of the Committee after consideration of the documentation before it, it is already possible to identify key questions that will require consultation with governments as well as with the international organizations concerned and the specialized commodity bodies. These issues include:

— specific stocking arrangements, and their techniques of operation—whether on a single-commodity or multi-commodity basis or both;

— refinement of estimates of the financial implications of international stocks, the size of the proposed fund, and the relation of its operational functions to sources of finance;

— proposals for multilateral contracts and compensation arrangements.

69. Besides the urgent need to endorse the general principles or key elements on the main lines of an integrated programme as outlined in the present report, so that further work on the programme can proceed without delay, the Committee could, in making its recommendations to the Board at its sixth special session give more concrete form and direction to the programme. In this regard the Committee might wish to recommend to the Board the setting up of suitable machinery and procedures to deal with specific issues such as those mentioned in paragraph 68 above.

70. The constitution and terms of reference of the machinery to be established in this respect should reflect the intention of bringing negotiated arrangements into force at the earliest possible time. It might take the form of a preparatory committee which would meet between the sixth special session and the fifteenth session of the Board to facilitate the taking of decisions on the programme at the latter session, decisions which would be aimed at negotiating without delay stocking arrangements, the establishment of a common fund and other aspects of the proposed programme. A preparatory committee that would be representative of the interests in the integrated programme could carry forward the elaboration of the proposals and allow the Board to give more adequate attention to the major policy decisions that would need to be taken. This might be the most practical procedure in view of the comprehensive character envisaged for the programme, particularly as the Board will have an unusually heavy agenda before it at its fifteenth session.

71. The Board at the first part of its fourteenth session expressed a sense of urgency in the matter of new approaches to commodity problems and policies and in particular, as already noted, regard-

ing the elaboration of the proposed integrated programme. It may also be observed that the World Food Conference, addressing itself to these matters in the context of the inter-relationship between the world food problem and international trade, made recommendations calling upon governments to devise, in appropriate organizations, effective steps for dealing with the problems of world markets, and, in urging UNCTAD to intensify its efforts in considering new approaches to international commodity problems, reiterated the recommendation of the Board in this respect to the Committee on Commodities. The World Food Conference also urged countries concerned and international financial institutions to give favourable consideration to the provision of adequate assistance to developing countries in cases of balance-of-payments difficulties arising from fluctuations in export receipts or import costs, particularly with regard to food.

72. In the light of the sense of urgency that is being shown by the international community on these issues, and the expectation that UNCTAD and other international organizations will move forward with expedition in deciding on the lines of the intergovernmental action required, the Committee will wish to focus its efforts on the recommendations to be made in this regard, and in particular on a time-table and programme of activity.

73. It should be borne in mind that other on-going activities between governments will be related to an integrated programme along the lines envisaged in the present report. These will include the multilateral trade negotiations with GATT, which are expected to move soon into a further stage of active negotiations and may be concerned with initiatives dealing with trade in commodities that would bear on aspects of the proposals for the integrated programme.

74. Specific financial questions relevant to the integrated programme proposals may also be taken up in 1975 in the agenda of work of the recently inaugurated Development Committee of the Governors of the World Bank and the IMF, and also by the Executive Boards of these agencies. In addition, there will be the need for re-negotiation of the international commodity agreements on tin and cocoa, as well as continuing efforts to re-establish economic provisions in other agreements or to negotiate new agreements. All of these activities will require close co-ordination with the over-all development of an integrated programme, in order that its basic principles and objectives, as well as the main emphasis to be given in the direction of its operation, should influence policy decisions elsewhere on aspects related to the programme and action on specific commodities.

75. Furthermore, it bears reiteration that the nature of the programme may well be imposed by world events. The gravity of the present international economic situation should persuade governments of the imperative need for early action on contingency policies, if the action eventually decided upon is not to be overtaken by more rapid change in the condition of the world economy, and of the developing countries in particular.

[1] General Assembly resolution 3202 (S—VI), Sect.I,3, para. (a), (iv).

[2] The General Assembly's Programme of Action called, in this respect, for "a link between the prices of exports of developing countries and the prices of their imports from developed countries" (Sect. I,1, para. (d) of resolution 3202 (S—VI)).

[3] Including the suitability of preferential terms for exports to developing countries where appropriate and technically feasible.

[4] The findings of UNCTAD studies on marketing and distribution of certain products will also be relevant to this objective. This programme of studies is being carried out in pursuance of Conference resolution 78 (III), and is the subject of a report to the Committee under agenda item 7.

[5] See TD/B/C.1/166/Supp.3 in which more formal multilateral contractual arrangements are considered.

[6] TD/B/C.1/166/Supp.1, paras. 41—43.

[7] Op. cit. para. 8.

[8] See also para. 71 below, for the views of the World Food Conference of relevance to this question.

[9] The complexity of the fats and oils group would entail careful examination of the feasibility of stock operations. Moreover the question of which particular coarse grains, and which particular vegetable oilseeds or oils, should be stocked would be a matter for further study. It might be possible to limit stock operations to certain key commodities in each group.

[10] Provided adequate facilities were available to them for the financing of stocks, exporting countries might not wish to conclude "symmetrical" contractual arrangements with importing countries, that is, arrangements incorporating a reciprocal commitment by the latter countries.

[11] See the report submitted by the UNCTAD secretariat to the intensive intergovernmental consultations on cotton held in April 1974 (TD/B/C.1/CONS.14/L.2, paras. 34 and 35).

[12] TD/B/C.1/166/Supp.4

Postscript: **THE SEVENTH SPECIAL SESSION OF THE UN GENERAL ASSEMBLY**

Ken Laidlaw

In June 1975, at the OECD-DAC High Level Meeting, the German delegate expressing his government's position on third world demands for a New International Economic Order, stated: "the advanced countries should be prepared to say clearly to the less developed countries that there are points on which they are not in agreement and that the electorate in developed countries is in general in favour of the existing system". At the Commonwealth Prime Ministers' Meeting in Jamaica a month earlier, Prime Minister Michael Manley of Jamaica emphasised that "at the heart of the problem of a new world economic order lie two issues. Firstly, how does the political leadership in the developed world convey to its internal political constituency the need for the kind of material accommodation which must be made if we are to commence the assault upon those wide disparities. Secondly, and perhaps this is more subtle, is the problem of political leadership accepting a redefinition of the exercise of sovereignty."

Clearly, an atmosphere of confrontation, incited by the rich world clinging to a belief in the present world economic system and the poor world's determination to create a new economic order, seemed inevitable before the 7th Special Session of the UN General Assembly in September 1975. The Session was called to consider development and international economic co-operation, a topic in which co-operation appeared to be far from a real possibility. Indeed, many observers felt that the UN as a mechanism for effective and co-operative action in international affairs, could well crumble in the face of protracted antagonism over the two week session.

In specific terms there were six areas of discussion put on the agenda for the delegates of 138 nations to discuss and negotiate. These were listed under international trade, transfer of real resources for development and international monetary reform, science and technology, industrialisation, food and agriculture and a restructuring of the United

Nations. Although it would appear difficult for the nations attending to arrive at a general concensus on such broad ranging topics it would not be from lack of preparation.

The industrialised nations for their part were becoming increasingly concerned over a number of issues evolving from the growing awareness of how interdependent the nations of the world were. Among these issues were:

(a) Continued recession: within this context it was becoming clear to most developed nations that their economic recovery was to some degree dependent on economic recovery among developing nations. Although these nations only import 15% of developed nations' manufactured exports, a cut back in imports would place an increasing strain on Western economic recovery.

(b) Political instability arising from present world economic difficulties.

(c) Increased oil prices: The OPEC nations were due to meet in Vienna two weeks after the session. Any reluctance on the part of industrialised nations to accept demands of Third World Nations could be countered by increased oil prices.

(d) Threat of producer cartels: Many developing countries would like to follow OPEC's example in which the advantages of "price taking rather than price making" were clearly shown. Although many studies, notably German, have shown that similar producer cartels are unlikely to succeed, developed nations would generally agree that united action by producers could disrupt western economies for a short period. This action under already existing economic problems would bring about an extremely difficult situation.

(e) Continued threat of Third World solidarity: Despite a number of problems facing individual countries, the anticipated dissolution of Third World solidarity has not taken place. The week before the Special Session a Conference of Non-aligned Nations in Lima reasserted this solidarity by agreeing to create a "solidarity fund" to be used to aid producer action by third world countries. Although this may have been a ploy by the poor nations in preparation for the Special Session the threat could not be ignored.

Despite these compelling reasons for co-operation in North/South negotiations some developed nations have been quite slow to respond, notably the U.S., Germany and Japan. The latter two are still reticent about the need to make concessions in improving relations with the Third World. However, by 1st September, the EEC had developed a common if somewhat nebulous approach to third world demands largely reminicent of the Lomé Convention, the U.S. had prepared what at first glance appeared to be a major policy revision in relations with the poor nations, the Commonwealth Finance Ministers had endorsed a far reaching report titled 'Towards a New Economic Order' (although Britain had voiced important reservations) and even Germany was beginning to waver on their original "the present economic order is just fine" attitude.

What about the developing countries? Their common approach in achieving development through complete economic emancipation had already been documented in three UN papers: The Declaration on the Establishment of a New International Economic Order, the Programme for Action, and the Charter of Economic Rights and Duties of States. Despite this apparent solidarity there were a number of factors which showed some cracks in third world armour and would no doubt be utilised by the rich nations at the session in order to soften their more radical demands. These included:

(a) Effects of oil prices on poor oil-consuming nations: although the oil-consuming countries of the Third World continued to voice support for OPEC, many were beginning to question the benefits of such action, especially the non-Muslim nations who were waiting in line for OPEC aid.

(b) Effects of economic recession in rich nations: despite attempts at OPEC type action by a number of commodity producers commodity prices continued to decline due to insufficient demand.

(c) The world food problem: it is the rich industrialised nations who produce the major portion of grains on the world market. Lack of purchasing power among the poor nations combined with a very tight supply/demand situation on the world market, has made them alarmingly dependent on the industrialised nations for a vital portion of their food supplies.

(d) The haves and have nots within OPEC: when all is said and done about OPEC financing buffer stocks in support of third world producer cartels, the fact of the matter is that two OPEC nations can determine its success. Both Iran and Saudi Arabia have yet to commit themselves to such schemes.

(e) Dependence on developed nations: the New International Order recognises continued third world dependence on the rich nations for financial, technological and industrial assistance.

Thus, the rich and poor nations meeting at the Special Session came together with the knowledge that each side possessed a certain number of levers which could be applied but at the same time there was an awareness of increasing interdependence between both parties. However, it should be acknowledged that the poor nations perhaps held the key, in that although they were aware that if muscles were to be flexed the consequences would be severe, they were also aware that if both sides were to lose, the rich nations had far more to lose than the poor. It was a question of confrontation or co-operation and the basis for negotiation would revolve around four working documents presented by the U.S., the Group of 77, the EEC and the Commonwealth.

After two weeks of intense negotiations the Special Session closed with the unanimous approval of a resolution outlining a series of trade and other measures to better the lot of the poorer countries. Indeed the Western press was euphoric in its assessment of the success of the Session and the degree of co-operation which was shown between the rich nations and the poor nations. The President of the Session, Foreign Minister Abdelaziz Bouteflika of Algeria, described it as an historic milestone. Many other Third World leaders were evidently pleased and certain that positive steps had been taken towards the creation of a new economic order. Daniel P Moynihan (U.S.) felt that the Special Session showed that genuine accord can be reached between the developed and developing nations and Ivor Richard (U.K.) said agreement had been reached because "the volume of demands was turned down and because compromises were offered and not simply called for from the other side".

Perhaps it is being overly cynical, especially *in lieu* of the statements above, to say that the Session was much rhetoric and little action. Indeed many consider the Session to be a huge success in which developed and developing nations discovered the ability to negotiate within a framework of a new sense of understanding and willingness to produce practical solutions. This theme of pragmatism was rolled out on the first day of the Session in Dr Kissinger's speech when he stated that "The United States has made its choice. There are no panaceas available — only challanges. The proposals that I shall announce today on behalf of President Ford, are a programme of practical steps, responding to the expressed concerns of developing countries. We have made a major effort to develop an agenda for effective international action; we are prepared in turn to consider the proposals of others. But the U.S. is committed to a constructive effort." One can't deny that a lot of practical solutions were put forth. In fact the depth and scope of Kissinger's solutions completely stunned many of the nations present for it represented an about-face in American policy towards the poorer nations. The U.S. has conceded that there is something wrong in the past relationships between the rich and poor world and something must be done about it. However, Kissinger's solutions will not bring about the "economic emancipation" called for by Third World nations. Instead they consist of more money to the IBRD and IMF for development aid, expansion of the international finance corporation and the creation of new organisations for technological and industrial research and information. There are a number of other proposals but they all amount to the same thing

— an attempt to provide new pillars to carry and maintain the present world economic system.

It is in this light that the final resolution must be viewed. Is it aimed at real structural change and the creation of new economic order which will bring about a more just and equitable distribution of the world's wealth? Or does it consist of a number of patches added to the present economic system — a system described by Mr J den Uyd (Prime Minister of Netherlands) as one in which "the growth of the gross world product over the last three decades has been enormous. But at the same time we have witnessed failures as a result of today's system — the uneven distribution of income between States and within countries . . . increasing destruction of our environment . . . and a threatening scarcity of resources; caused by unlimited exploitation".

In terms of actual results aimed at fundamental change the Resolution provides no reason for joy among those seriously concerned about the creation of a more just and equitable world economic system. Admittedly, there was a good deal of agreement combined with a number of detailed steps to help the Third World countries with science and technology, to spur their industrial and agricultural development, to provide greater access to industrialised markets and a greater voice in international monetary reform. There was also agreement by the OECD countries (with the exception of the U.S. and Britain) to re-commit themselves to 0.7% of their GNP by 1980 as official development assistance.

However, in the controversial area calling for a restructuring of the international trading system by means of integrated commodity agreements, buffer stocks and a common fund to finance them, multilateral commitments and indexation of the price of the commodity to the price of manufactured goods, there were a number of serious reservations put forth by the developed nations which indicate that fundamental change is not what they seek. In other words, the resolution, conceptually may provide the framework for a new economic order, but realistically there is still a wide gap between the developed and developing nations in terms of their political will to create a new order. As Mr Jacob Myerson, the U.S. representative put it, "the U.S. cannot, and does not accept any implication that the world is now embarked on the establishment of something called 'the new international economic order' ".

The final resolution asks developed countries to continue to study the area concerned with international commodity agreements, buffer stocks and a common fund to finance them. The implementation of such schemes are vital to the NIEO. However, the U.S. specifically dissociated itself from any moral commitment to give serious consideration to fundamental change in the present 'free-market' system. In the key area of indexation the U.S. agrees to join others in a study but at the same time qualifies such accord with the statement that "the U.S. has to make it clear that it does not support such a proposal". Britain has also stated that she is against indexation. Such views may meet with opposition from less developed countries who saw the price of their imports — the large majority consisting of manufactured goods — rise 40% while their export prices increased only 27%. It is probable that the non-oil exporting members of the Third World will see the purchasing power of exports of primary products fall by 13% during 1975.

To conclude, the Special Session should be seen as a minor skirmish with the real encounter yet to come. Albeit, there has been a perceptible change in the attitude of the developed nation towards the developing and this must be seen as a positive step towards the fundamental change needed. However, it is wishful thinking to suppose that the resolution indicates everything to be well. Solutions to global poverty are unlikely to be found until developed nations are prepared to accept the need for a fundamentally restructured economic system. Until this comes about the present system will continue to work against the interests of 70% of the globe's inhabitants.

APPENDIX

Notes on Contributors

F E Banks

Dr Fred Banks is a research fellow in economics and econometrics at the University of Uppsala, Sweden. He has taught at the University of Stockholm, the UN African Institute for Economic and Development Planning, Dakar, Senegal and has been OECD Lecturer in Macroeconomics in Lisbon. He is author of "The World Copper Market: an Economic Analysis" and is at present working on another book under a Ford Foundation grant.

Biplab Disgupta

Dr Biplab Dasgupta studied economics at the Universities of Calcutta and London and also has an MSc in Computer Science. He was a lecturer in the School of Oriental and African Studies at the University of London from 1969 to 1972 and since then has been a Fellow at the Institute of Development Studies at the University of Sussex. He is the author of five books including one on the oil industry in India and is currently working on three further books on aspects of rural development.

Robert Dickson

Bob Dickson studied at the Universities of Glasgow and Aberdeen and has an MA in Classics. He finds this background surprisingly useful for his present job as North Field Officer for the World Development Movement, a British development lobby with some 200 local branches. He has worked for WDM for 2½ years and has developed an extensive interest in political aspects of international development, especially in connection with relations between developed and less developed countries.

Paul Rogers

Dr Paul Rogers studied at Imperial College, London from 1961 to 1967 and then spent three years working for the Overseas Development Ministry, two of them on a regional crop improvement programme in East Africa. Since 1971 he has lectured at Huddersfield Polytechnic and is involved in the new Human Ecology Degree Programme there. He is interested in the political aspects of resource use and has written a number of articles and papers on this theme.

Anthony Vann

Anthony Vann took a degree in Botany at the University College of Wales at Aberystwyth and then did research for three years on grassland ecology. He has lectured at Huddersfield Polytechnic since 1971 and his main teaching and research interests are in plant ecology. He is also involved in the Human Ecology Degree Programme especially in components on resource studies and ecology. He and Paul Rogers collaborated on a previous book in this series, "Human Ecology and World Development", the proceedings of a symposium held at the Polytechnic in April 1973.

Ken Laidlaw

Although Scottish by birth, Ken Laidlaw was educated in Canada, taking a degree in Political Science and History at the University of Waterloo, Ontario. After postgraduate studies in political science he worked for two Canadian development organisations, the Development Education Centre in Toronto and Gatt-Fly, the latter being an inter-church lobbying and educational group. He now works as Information Officer with the World Development Movement in London and is currently researching issues connected with the New International Economic Order and the fourth session of the UN Conference on Trade and Development.

INDEX